REPAIR, PROTECTION AND WATERPROOFING
OF CONCRETE STRUCTURES

REPAIR, PROTECTION
AND WATERPROOFING
OF CONCRETE STRUCTURES

PHILIP H. PERKINS

CEng,FICE,FASCE,FCIArb,FIPHE,MIWES

ELSEVIER APPLIED SCIENCE PUBLISHERS
LONDON and NEW YORK

ELSEVIER APPLIED SCIENCE PUBLISHERS LTD
Crown House, Linton Road, Barking, Essex IG11 8JU, England

Sole Distributor in the USA and Canada
ELSEVIER SCIENCE PUBLISHING CO., INC.
52 Vanderbilt Avenue, New York, NY 10017, USA

WITH 3 TABLES AND 76 ILLUSTRATIONS

© ELSEVIER APPLIED SCIENCE PUBLISHERS LTD 1986

British Library Cataloguing in Publication Data

Perkins, Philip H.
Repair, protection and waterproofing of
concrete structures.
1. Reinforced concrete construction —
Defects   2. Buildings — Repair and
reconstruction
I. Title
624.1′8341   TA683.2

Library of Congress Cataloging in Publication Data

Perkins, Philip Harold.
Repair, protection, and waterproofing of concrete
structures.

Updated ed. of: Concrete structures. 1977.
Bibliography: p.
Includes index.
1. Concrete construction—Maintenance and repair.
2. Waterproofing. 3. Concrete coatings. I. Perkins,
Philip Harold. Concrete structures. II. Title.
TA681.P42 1986   624′.1834′028   86–11644

ISBN 1-85166-008-9

Typeset by Interprint Malta Ltd.

Printed in Great Britain by Galliard (Printers) Ltd, Great Yarmouth.

Preface

Since the author's first book on the repair, waterproofing and protection of concrete structures was published in 1976, the need for repairs to this type of structure has increased dramatically.

It has been estimated that in 1982, the value of repairs and maintenance to buildings in the UK amounted to some $£8 \times 10^9$ and if civil engineering structures were included this would rise to $£1 \times 10^{10}$ (£10 000 million). These figures relate to all types of buildings and structures of which concrete forms only a part. The disturbing feature is that in most cases the structures are not more than about 25 years old. The problem is not confined to the United Kingdom.

While the structural design of reinforced concrete varies to some extent from one country to another, the materials used, Portland cement, aggregates and steel reinforcement, are essentially similar. The causes underlying the deterioration are basically the same in all countries and the principles involved in dealing with the deterioration are also similar.

This book is intended to deal mainly with 'non-structural' repairs, that is repairs which are intended to restore long-term durability, but which will not increase to any significant degree the load bearing capacity of the structure. Mention is made of certain aspects of the execution of structural repairs but the calculations necessary for the design are not included. However, one of the first and most important steps in the investigation of a deteriorated structure is to decide whether structural strengthening of the structure is needed. For this reason the author recommends that all such investigations should be carried out by a Chartered Civil or Structural Engineer with considerable experience in this type of work. It should be remembered that the investigation and diagnosis and subsequent preparation of specification for remedial work to a deteriorated structure is quite different to the design of a new structure.

Reinforced concrete structures which have been properly designed and constructed and which operate under normal conditions of exposure and use, require only a minimum of maintenance. However, it is a fundamental error to assume that they are maintenance free. Regular and careful inspection and maintenance are essential for all structures.

It is important that the lessons which can be learnt from the correct diagnosis of the present problems should be put to practical use in the design and construction of new structures.

The author wishes to express his gratitude to the many professional men from whom he received help and information. Special thanks are due to Keith Green of Burks, Green & Partners, Consulting Engineers, and George Korab of the Cement Gun Group of Companies for many informative discussions. Mention must also be made of the work of Lucy Perkins for reading and checking the manuscript.

PHILIP H. PERKINS

Contents

Introduction

The author considers that while no structure is maintenance free, large scale deterioration should not occur. The obvious exception to this is damage caused by fire, impact, explosion, etc. Structures in a hostile environment and/or subjected to severe conditions of use, should, as far as possible, be kept under close observation and subject to regular inspections. One example of this is the oil production structures in the North Sea, where the most elaborate precautions are taken; a detailed discussion of these is outside the scope of this book.

The use of reinforced concrete as a major constructional material increased enormously all over the world after the end of the Second World War. The amount of concrete and the speed of design and construction was quite phenomenal. The author's experience suggests that this is one of the basic causes of trouble which has developed and which is likely to increase for many years to come.

An essential feature in the long term durability of a reinforced concrete structure is the protection of the steel reinforcement. A very high percentage of the deterioration which has occurred arises from failure to provide this protection. It has been found that a percentage of chlorides above a certain level in the concrete can result in serious corrosion of the steel rebars and this is discussed in some detail in this book. In recent years there has been much concern in many countries, including the United Kingdom, arising from the deterioration of the concrete itself by alkali–aggregate reaction. In the UK in 1971 there was one reported case but by 1985 some sixty cases had been confirmed by the Cement and Concrete Association. However, the *New Civil Engineer*, in May 1985, estimated that the number of suspected cases of ASR (alkali–silica reaction) in road bridges was about 350. Repair techniques are available which, when properly carried out, will mitigate the effects and slow down the rate of deterioration and these are discussed in the book.

The author recommends that for repairs of any magnitude the following basic procedure should be adopted:

The appointment, by the building owner, of an experienced Chartered Engineer (Civil or Structural), who would then be responsible for the following:

(a) Deciding on the details of the investigation and instructing a commercial testing laboratory.
(b) Diagnosis of the causes of the deterioration.
(c) Preparation of report to the client.
(d) Preparation of specification and contract documents.
(e) Advising on the adjudication of the contract.
(f) Inspection/supervision of the work.
(g) Regular post contract inspection and monitoring and advising on a practical programme of maintenance.

Chapter 1

The Principal Materials Used in the Repair of All Types of Concrete Structures

1.1 INTRODUCTION

The range of materials which can be effectively used for the durable repair of concrete structures is fairly limited. Those most widely used are concrete and mortar, made as far as practicable with the same type of cement and aggregates as were used in the original structure. When deterioration is due to chemical attack, it may be necessary to use a different cement and/or protective coatings.

When repaired areas fail it is in many cases due to failure or partial failure of the bond between the old and new work. The standard of bond developed between the old and new concrete is directly related to the care taken in the preparation of the base concrete. In recent years a great deal of attention has been paid to the development of bonding agents.

The vast majority of concrete structures which need repair are reinforced and the corrosion of the reinforcement plays a decisive part in the deterioration of the structure. When ferrous metals corrode, the corrosion products occupy a larger volume than the original metal and the resulting expansion causes disintegration (spalling) of part of the surrounding concrete.

The steel passivating properties of the original concrete are of great importance in providing long term durability to the reinforcement and to the structure as a whole. Good quality Portland cement concrete and mortar can provide permanent protection to the steel due largely to the alkaline environment created by the cement paste. This important matter is dealt with in some detail in Chapter 2.

The selection of suitable materials to replace defective and spalled concrete and to reintroduce a protective and durable environment around the reinforcement is of great importance.

1

1.2 CEMENTS

For the purpose of this book cements are classified as Portland cements and non-Portland cements.

The quantity of Portland cement used in the construction industry in the UK far exceeds that of all other types and in 1986 amounted to about 13 million tonnes.

1.2.1 Portland Cement: Ordinary and Rapid Hardening

In the UK these two cements are covered by British Standard BS 12 and form the bulk of the Portland cements used for the repair of concrete structures, with ordinary Portland predominating.

The basic difference between the two cements is the rate of gain of strength. The increase with the rapid hardening cement is largely due to the finer grinding and this cement usually has a specific surface of about 4300 cm^2/g. The rapid hardening is accompanied by an increase in the rate of evolution of heat of hydration which in turn raises the temperature of the maturing concrete during the first 15–40 h after casting. It should be noted, however, that the rate of hardening and evolution of heat of hydration does not depend on the fineness of grinding alone, and that the chemical composition of the cement also plays a part.

1.2.2 Sulphate-Resisting Portland Cement

The relevant British Standard is BS 4027 and the cement is similar in its strength and other physical properties to ordinary Portland cement. However, it is generally darker in colour than most ordinary and rapid hardening Portland cements. The essential difference is in the limitation of the tricalcium aluminate (C_3A) content to a maximum of 3%.

It is the tricalcium aluminate in Portland cement which is attacked by sulphates in solution and this chemical reaction results in the formation of ettringite which can have a disruptive effect on the concrete, causing dimensional changes and reduction in strength. It is advisable to consult the cement manufacturer before using any type of admixture with this type of cement, but calcium chloride should not be used as the sulphate resistance in the long term will be reduced.

Sulphate-resisting Portland cement, in common with all Portland cements, is vulnerable to acid attack.

1.2.3 White and Coloured Portland Cements

White Portland cement is a true Portland cement and complies with BS

12 (Portland cement: ordinary and rapid hardening). The special point about this cement is that the raw materials are specially selected; the clay is a white china clay and the manganese and iron content is kept to an absolute minimum.

Coloured Portland cements, other than white and the pastel shades, generally consist of ordinary Portland with a pigment ground in at the works. The pigments used are covered by BS 1014—Pigments for cement and concrete. Coloured Portland cements are now included in BS 12.

These cements would only be used in cases where a colour match with the existing concrete was desirable. However, due to colour changes with age and weathering, it is likely that the best results would be obtained by using pigments and trial mixes.

A question is sometimes raised as to whether it is necessary to increase the cement content of a concrete or mortar mix to allow for the presence of the pigment in the cement. It is not possible to give an answer to this in the form of a straight 'yes' or 'no'. Where durability and/or impermeability are important factors, the author considers that allowance for the pigment should be made by increasing the specified cement content by weight of pigment. Where strength is overriding, then the only satisfactory solution is to make trial mixes because the strength of ordinary Portland cement varies within certain limits.

At the time of writing this book, coloured (pigmented) Portland cement was no longer available in small quantities on the UK market.

Other Portland type cements which are used to a limited extent in construction are Portland Blast-furnace cement (BS 146) and Portland Blast-furnace Low Heat cement (BS 4246), but the author has not come across these cements being used for repair work.

In 1985, two new British Standards were published for what may be termed 'blended cements'. These are:

Portland Pulverised Fuel Ash cement (BS 6588), and
Pozzolanic Cement (with pulverised fuel ash as pozzolana) (BS 6610)

In BS 6588, the PFA content is limited to a maximum of 35% and a minimum of 15% by mass of the total quantity. In BS 6610, the limits for PFA content are, maximum 50% and minimum 30%. In both Standards the PFA can be either ground-in or mixed on site in the mixer. There is no requirement for pre-bagging, so that if site mixing is adopted, sampling would have to take place at the mixer and prior to the addition of water and aggregate. Once the concrete or mortar is mixed, it is no longer possible to check on the proportion of PFA in the mix.

Some authorities are very concerned at this apparent departure from a basic principle of British Standards for materials, namely that there must be proper and readily available opportunity for sampling and testing to check compliance prior to use. The fact is that both these cement Standards require that sampling and testing shall be carried out in accordance with BS 4550:Part 1—Methods of Testing Cement—Sampling. Clauses 1 and 5 of this Standard lay down the 'sampling situations' and the methods of sampling to be used for taking samples in these situations. It is doubtful if, from a practical point of view, compliance with BS 4550:Part 1 could be obtained when the two cements are made by 'dry blending' outside a cement factory. To comply with the Standards, these two cements must be placed in a 'sampling situation' which is in accordance with BS 4550:Part 1.

1.2.4 High Alumina Cement (HAC)
High alumina cement is covered by BS 915. About 60% of the world consumption of this cement (outside the USSR) is produced by Lafarge Fondu International.

HAC differs fundamentally from Portland cement as it consists predominantly of calcium aluminates. It is much darker in colour than ordinary and rapid hardening Portland cement. The lighter shades of HAC and the darker shades of sulphate-resisting Portland cement may approach each other in colour. The setting time is similar to that of Portland cement. With increase in ambient temperature the setting time tends to increase whereas the reverse is the case with Portland cement. However, with temperatures above with 30°C, the setting time is likely to be reduced and in extreme cases there may even be a 'flash' set.

The rate of gain of strength of HAC is very rapid, normally reaching about 80% of its nominal maximum in 24 h. This rapid increase in strength is accompanied by a rapid evolution of heat of hydration. This has advantages and disadvantages; it is very useful when working in low temperatures, but very careful wet curing is required to prevent thermal cracking. The rapid gain in strength makes it extremely useful for emergency repairs.

HAC concrete which has been correctly proportioned, placed, compacted and cured, exhibits improved resistance to many chemical compounds, including sulphates, sugars, vegetable oils and dilute acids, compared with Portland cement concrete. On the other hand it is vulnerable to attack by alkalis and chlorides.

To achieve long term high strength, adequate durability and maximum chemical resistance, the cement content should not be less than $400 \, kg/m^3$ and the water/cement ratio should not exceed 0·4; a lower water/cement ratio is preferred. In addition, thorough compaction and careful water curing is required. Admixtures should only be used with the approval of the manufacturers; by the use of special accelerators recommended by the cement makers, setting time can be reduced to a few minutes, but this is accompanied by a reduction in strength.

Since the middle of the 1970s high alumina cement has not been permitted to be used in the UK for structural concrete under the National Building Regulations and relevant Codes of Practice. While in the past there were some restrictions on the structural use of this cement in France, these have now been lifted provided the high alumina cement is made by French Fondu Lafarge, and the concrete complies with specific requirements. Detailed advice and information on HAC can be obtained from the manufacturers, Lafarge Fondu International.

1.2.5 Chemically Resistant Cements

These special cements are not used for concrete except where very small quantities are required. They are used for mortar and grout for bedding and jointing chemically resistant tiles and bricks.

The two basic types of cements are resin cements and silicate cements. The resin type cements are mainly based on modified phenolic resins, blended epoxies, cashew nut resin cement, furane resin, and polyester resin. The silicate cements are based on sodium silicate and potassium silicate, and are resistant to very high temperatures as well as a wide range of aggressive chemicals.

Chemically resistant cements usually consist of a powder and a gauging liquid (sometimes termed a syrup) which are mixed together in the prescribed proportions immediately before use.

Detailed information on these cements is contained in a small handbook issued by Prodorite Ltd of Wednesbury, included in the Bibliography at the end of this chapter.

1.3 COMMENTS AND COMPARISONS OF PORTLAND CEMENTS FROM THE UK AND SOME OTHER COUNTRIES

1.3.1 General

The direct comparison of the properties of Portland cements made in different countries to the relevant National Standards is very difficult. In international contracting it is often found that cement which is available in the country where the project is to be built originates in a country other than that where the specifier has personal knowledge of the relevant National Standard. In dealing with such a situation, the specifier should consider not only the requirements in the relevant National Standard but also, as far as this is possible, the performance of the concrete made from the cement in the country of origin.

As previously stated, Portland cement is produced and used in far greater quantities for construction work than any other type of cement. It is therefore felt that some general information on the terms used internationally would be useful.

Portland cement is usually designated and classified as follows:

OC. Ordinary Portland cement.
RHC. Rapid hardening Portland cement, or High Early Strength cement or High Initial Strength Portland cement.
HSC. High Strength Portland cement.
LHC. Low Heat Portland cement.
SRC. Sulphate-Resisting Portland cement.
AEC. Air-Entrained Portland cement.

It is assumed for the purpose of chemical analysis of Portland cement based mortars and concretes, that Portland cement contains about 64% calcium oxide (CaO). It should be noted that in the hydrated cement paste this is present in he form of complex hydrates.

The principal consituents of Portland cement are usually expressed as follows:

Tricalcium silicate	(C_3S)
Dicalcium silicate	(C_2S)
Tricalcium aluminate	(C_3A)
Tetracalcium alumino ferrite	(C_4AF)

A direct comparison of National Standard Specifications of one country with Standards for similar materials in another country can be misleading. The test requirements and methods of test are different. For example, if it is considered essential that a cement should comply in all respects with the requirements of BS 4027—Sulphate-resisting Portland cement, then the only reasonable solution is to have the cement tested by the methods applicable to BS 4027.

The author would however point out that failure of a cement made outside the UK to comply with all the test requirements in the relevant

BS would not necessarily mean that the cement was not suitable for the proposed use.

Questions are often asked about comparison of certain US cements and UK cements, particularly those with sulphate-resisting properties. For example, sulphate-resisting low heat Portland cement to ASTM Standard C 150–84, Type II, has no real equivalent in the UK. It is a moderately resistant Portland cement, which perhaps could be compared with cement to BS 6588—Portland Pulverised Fuel Ash cement. Cement to ASTM C 150–84, Type V, is a sulphate-resisting Portland cement in which the C_3A content is limited to 5%, while in cement to BS 4027, the C_3A content is limited to 3%.

In some of the Continental countries, such as Austria, France and West Germany, the denomination and type of cement is expressed by capital letters followed by a number of digits; for example, PZ 35–45 would denote a Portland cement with a 28 day compressive strength of 35–45 N/mm^2 (DIN 1164–1978) when tested according to the National Standard.

One aspect of Portland cements in the UK requires special comment, namely, the concern felt by many concrete engineers that important changes have occurred in the properties of Portland cement (ordinary and rapid hardening) over the past 25 years or so. A Working Party of the Concrete Society produced a Report on this in October 1984; those interested should read the Report itself (see Bibliography at the end of this chapter).

Very briefly, the following important facts emerged:

(a) There had been no significant change in particle size.
(b) The overall proportion of silicates had remained virtually unchanged, but the ratio of tricalcium silicate to dicalcium silicate had risen significantly.
(c) The overall proportions of other compounds had remained reasonably constant.
(d) There had been a marked increase in early age strength.
(e) The long term increase in strength after 28 days had been reduced, so that now it was probably justified in most cases not to consider any such increase.

The general conclusion was that it was essential not to specify concrete by strength alone if long term durability was important. Minimum cement content and maximum water/cement ratio were very important.

1.4 STEEL REINFORCEMENT

1.4.1 Mild and High Tensile Steel

Steel reinforcement for concrete is covered by the following British Standards:

BS 4449—Hot rolled steel bars for reinforcement of concrete.
BS 4461—Cold worked steel bars for reinforcement of concrete.
BS 4482—Hard drawn mild steel wire for reinforcement of concrete.
BS 4483—Steel fabric for reinforcement of concrete.

There is one British Standard for prestressing wire and strand, namely BS 5896.

The coefficient of expansion of plain carbon steel is 12×10^{-6}.

1.4.2 Galvanised Reinforcement

At the time of writing this book there is no British Standard for galvanised reinforcement. Galvanised rebars and prestressing wire are used to a limited extent in the UK and to a much greater extent in the USA.

The object of the galvanising is to provide additional protection to the rebars and prestressing wire in especially adverse conditions of exposure. Galvanising consists of coating the steel with zinc, either by dipping the steel into tanks of molten zinc or by electrodeposition from an aqueous solution.

The principal British Standard for galvanising steel is BS 729—Hot-dip galvanised coating on iron and steel articles. The thickness of the coating has an essential influence on durability, and very useful information on the subject can be obtained from the Zinc Development Association, 34 Berkeley Square, London, who have published a Galvanising Guide. In the USA, the International Lead Zinc Research Organization of 292 Madison Avenue, New York, have issued an excellent book, 'Galvanised Reinforcement for Concrete', Vol. 2.

Because the galvanising is done with zinc, there is a chemical reaction between the caustic alkalis in the cement paste and the zinc, with the evolution of hydrogen. Generally, this has no ill effects, but if the repair is carried out with mortar there is a slight risk that some debonding may occur, particularly on soffits and vertical faces. To avoid this, the galvanised steel can be given a chromate wash which will inhibit the chemical reaction and thus prevent the evolution of the hydrogen.

1.4.3 Epoxy Powder Coated Reinforcement

This was introduced in the US and Canada in the mid-1970s but only arrived in the UK in the early 1980s. The epoxy powder is electrostatically applied, and in the US the coated bars have to pass the tests laid down in the Standards of ASTM and the US Federal Highway Authority, and are used mainly for bridge decks and bridge abutments.

The principal US Standard is ASTM No. A 775–81: Epoxy-coated Reinforcing Bars, which lays down basic requirements, but a number of other ASTM Standards are involved as well as those of the National Association of Corrosion Engineers.

As far as the UK is concerned, there is still considerable resistance to the use of any protective coatings on steel reinforcement in concrete on the grounds that a properly designed concrete mix which is correctly placed, compacted and cured, with the provision of adequate cover to the steel, will ensure very long term protection to the reinforcement. There is no doubt that technically, this is correct, but the large number of defects in reinforced concrete structures of all types which have been caused by the corrosion of the reinforcement suggests that in the past far too much reliance has been placed on the standard of site workmanship. This aspect of defects in reinforced concrete structures is discussed in more detail later in this book.

At the time of writing this book, in the US the extra cost of epoxy powder coated rebars over uncoated bars is about 15–20%. In the UK the increase is greater due to the very small demand.

1.4.4 Stainless Steel

There are three basic groups of stainless steel: martensitic, ferritic and austenitic. Of these, it is the austenitic steels which are the most widely used in building and engineering and the information which follows relates to this group. Austenitic steel is an alloy of iron, chromium and nickel, and two types in this group contain also a small percentage of molybdenum. The type most generally suitable for use in external repair work in En58J (also-known as 316 steel), which contains 18% chromium, 10% nickel and 3% molybdenum. It is very resistant to corrosion but is also very expensive compared with mild steel. The steel can be welded. However, it may be non-magnetic or only slightly magnetic, and therefore it may not be possible to locate it with a normal type of cover meter. A further point is that mild steel in contact with stainless steel may suffer accelerated corrosion, i.e. it may be anodic to stainless steel.

This steel (En58J) is covered by **BS 1449:Part 2**.

The use of stainless steel is only justified in cases where protection of normal steel cannot be relied upon due to difficulty of application or difficulty of maintenance of the protective treatment. In repair work its use is likely to be confined to pins, fixings and anchors, and stainless steel mesh for reinforcing relatively small areas which are repaired with mortar.

The coefficient of expansion of austenitic stainless steel is 18×10^{-6}, that of ferritic stainless steel is 10×10^{-6}.

1.5 NON-FERROUS METALS

A limited number of non-ferrous metals are used in repair work; when these are used, it is hoped that the following information and recommendations will prove useful. Reference should also be made to Chapter 2, Sections 2.4.4 and 2.4.5.

1.5.1 Aluminium

If unprotected by anodising or suitable coatings, aluminium in direct contact with damp concrete may be attacked by the caustic alkalis in the cement. Modern methods of protection include anodising, stoved polyester powder and stoved pigmented paints. Further information on this subject can be obtained from **BS 4873**—Aluminium alloy windows. Specialist advice should be sought if aluminium and steel are in direct contact.

The coefficient of thermal expansion of aluminium is about $25 \times 10^{-6}/°C$.

1.5.2 Copper

Copper is resistant to most conditions met in building construction and water and sewerage works. It is not corroded by Portland cement concrete unless chlorides are present. Ammonium compounds may attack copper. If copper and steel are in contact in the presence of moisture, there is the danger that steel will corrode as it is anodic to copper.

The coefficient of thermal expansion of copper is about $17 \times 10^{-6}/°C$.

Copper used to be specified for water bars, but has been largely replaced by PVC.

1.5.3 Phosphor-Bronze and Gunmetal

Bronze is an alloy of copper and tin, and phosphor-bronze contains phosphorus as copper phosphide. It is used in conditions which would result in the corrosion of ferrous metals, mostly for fixtures and fittings. The same comments apply to gunmetal which is bronze with about 9% of tin.

The coefficient of thermal expansion of phosphor-bronze is about 20×10^{-6}.

1.6 AGGREGATES

For repairs, aggregates for concrete and mortar are generally considered less important than for new structures. Nevertheless, it is felt that some general remarks may be useful. In the UK, aggregates obtained from 'natural sources', such as gravels, pit sand, quarried rock and sea-dredged material, should comply with BS 882—Aggregates from natural sources for concrete. Tests on aggregates are covered by BS 812—Methods for sampling and testing of mineral aggregates, sands and fillers. It is important to note that the British Standard 882 contains the following statement:

> 'No simple tests for the durability and frost resistance of concrete or for corrosion of reinforcement can be applied and experience of the properties of concrete made with the type of aggregates in question and a knowledge of their source are the only reliable means of assessment....'

Shrinkage of natural aggregates, when excessive, can create problems. In the UK this characteristic has so far only appeared in certain aggregates in Scotland, but in other parts of the world it has proved to be a significant factor in the selection of aggregate sources. This characteristic is dealt with in detail in the Building Research Establishment's Digest No. 35—Shrinkage of Natural Aggregates in Concrete. Aggregate–alkali reaction is another matter which should be considered when dealing with aggregates from a previously unknown or untried source. This particular problem is extremely complicated and is discussed later in this book.

Matters which can be important, depending on the circumstances of each case, include the following:

(a) Source and classification.

(b) Particle shape and surface texture.
(c) Grading (sieve analysis) including clay, silt and dust content.
(d) Organic impurities.
(e) Salt content, particularly of the fine aggregate, with particular reference to chlorides and sulphates.
(f) Mechanical properties.
(g) Flakiness index.
(h) Shell content; this applies principally to sea-dredged aggregates.

Brief comments are given below on items (d), (e), (g) and (h).

(d) *Organic impurities.* It has been found very difficult to define and place limits on organic impurities. The most satisfactory procedure is to carry out tests on the 'suspect' aggregates and compare results with concrete or mortar made with standard aggregates. The usual tests would be setting time, rate of gain of early strength, and 7 and 28 day strengths. One contaminant sometimes found in pit gravels in certain parts of the UK is coal (lignite). A practical limit for this is 0·5% by weight of the aggregate. An aggregate known to contain coal should not be used for concrete for floors which are frequently washed down or subjected to heavy abrasion.

(e) *Salts—chlorides and sulphates.* The presence of these impurities has caused serious deterioration of concrete in the Middle East and other parts of the world. In the UK the chloride content of concrete is strictly controlled by the recommendations in Code of Practice BS 8110. Reference should be made to the Code for details, but for practical purposes, chloride ion concentration in reinforced concrete made with OPC or RHPC is limited to 0·40% by weight of cement. The chloride ion content of the aggregates on their own is limited by Appendix C of BS 882 to 0·06% by weight of total aggregate for 95% of test results, with no result exceeding 0·08%, for reinforced concrete made with OPC or RHPC. Generally it is marine dredged aggregates which contain chlorides and these need careful washing.

The sulphate content of aggregates is not mentioned in BS 882 because it has not been found to be a problem in the UK. The sulphate content of sea water is about 10% of the chloride content (i.e. about 2500 ppm or mg/litre). A generally accepted limit for use in countries where there is a problem, is about 4·5% by weight of ordinary Portland cement and 6% by weight of sulphate resisting Portland cement. Portland cements contain gypsum to control the set (about 2·0–3·0% by weight). The objection to sulphates is that they react with the C_3A in Portland cement to form ettringite, a reaction which is expansive in character and weakens or disintegrates the concrete. Sulphate attack on concrete is discussed later in this book. See also Section 2.5 in Chapter 2. Limits on sulphate content of certain aggregates are given in the relevant British Standards, i.e. BS 877, BS 1047 and BS 3797.

(g) *Flakiness index.* This is covered by clause 4.2 in BS 882. The effect of a large number of flakey particles in an aggregate is to reduce workability with a given w/c ratio and to reduce abrasion resistance. It can be important in external paving, floors, and marine structures subject to abrasion from shingle and sand.

(h) *Shell content.* This is covered by clause 4.3 in BS 882, and as the effect of shell in concrete is often misunderstood, Table 2 of BS 882 is reproduced below in abbreviated form.

Aggregate size (mm)	Maximum shell content (% of dry weight)
40 ⎱ 20 ⎰	8
10	20
Fine aggregate	No requirement

The effect of the shell content depends on the size of the shells and their shape and the environmental conditions (exposure) under which the concrete will have to function. Large shells can adversely affect durability, and can be detrimental to the appearance of fair-faced concrete. Poorly shaped fine shell can increase the water demand of the mix. As shell is mostly calcareous, this must be allowed for in chemical analysis of concrete to determine cement content, in the same way as for limestone aggregate.

1.7 ADMIXTURES FOR CONCRETE

A simple definition of an admixture is that it is a chemical compound which is added to concrete, mortar or grout at the time of mixing for the purpose of imparting some additional and desirable characteristic(s) to the mix.

Admixtures are sometimes referred to as 'additives', but it is better to use the latter word for the addition of chemical compounds to cement at the cement works, where they are ground-in at the time of manufacture.

Admixtures should only be used when they are really required to produce a particular result which cannot be obtained by normal mix design.

Admixtures should not be used with any cement except ordinary and rapid hardening Portland cement without the approval of the cement manufacturer.

The following are the main purposes for which admixtures are used:

(a) To accelerate the setting of the cement and the hardening of the concrete, mortar or grout; these compounds are known as accelerators.
(b) To retard the setting of the cement and slow down the rate of hardening of the mix. These are known as retarders.
(c) To entrain air in the mix (air-entraining agents). These compounds give an air-entrained mix, which should not be confused with an aerated mix. The latter is obtained by quite different compounds and is used for different purposes.
(d) As expanding agents in mortar and grout to neutralise the effect of drying shrinkage.

1.7.1 Accelerators

Accelerators can be useful in cold weather and urgent repair work, such as work between the tides and patching of floors.

The great majority of accelerators used in concrete are based on calcium chloride ($CaCl_2$) as the active ingredient. The use of this compound, apart from speeding up the chemical reaction of cement and water, has certain other effects, the most important of which are:

(a) Calcium chloride is very aggressive to ferrous metals.
(b) It increases drying shrinkage of the mix.
(c) It reduces the sulphate resistance of sulphate-resisting Portland cement.

Because of the serious disadvantages of calcium chloride mentioned above, considerable efforts have been made to find a satisfactory substitute as the basis for accelerators. So far only two compounds have met with even a limited degree of practical success, and these are calcium formate and sodium carbonate.

The most effective and satisfactory method of speeding up the setting and hardening of Portland cement concrete is the use of heated concrete or the application of heat to the concrete after casting.

1.7.2 Retarders

There are two main uses for retarders. One is as an integral part of a mortar or grout or concrete mix when it is required to extend the setting time of the cement and reduce the rate of hardening of the concrete. The other is when the retarder is used on formwork to retard the setting and hardening of the surface only of the concrete, in order to facilitate work on the concrete surface when the formwork is removed.

Retarders are usually sugars and similar compounds, but borax is also used. The reaction between retarders and Portland cement is a very complicated one. It is affected by the chemical composition of the cement and the temperature of the concrete as it is maturing. Therefore the period of retardation can only be estimated approximately and accurate reproduction of results is difficult and requires considerable experience.

1.7.3 Air-Entraining Agents

Air-entrained concrete is used for roads and external pavings, to resist the disintegrating effects of frost and de-icing salts. Air-entraining agents, however, can be useful for concrete repair work and for mortar used on repairs to structures on very exposed sites in the northern parts of the UK, as well as for marine structures in northern latitudes.

While these compounds were originally introduced to provide a concrete which would resist frost action, it has been found that they impart other beneficial characteristics to the mix. These are:

(a) They help to reduce, and may, in favourable circumstances, eliminate plastic cracking.
(b) They help to reduce a tendency to water scour on the surface of fair-faced concrete and reduce segregation.
(c) They improve workability.

The best air-entraining agents are resins. They must be carefully dispensed into the mixing water, so that the dosage is accurately controlled and the compound is uniformly distributed throughout each batch. Their effect is to produce a large quantity of minute bubbles of air which alter the pore structure of the concrete. The effect of this entrainment of air (about $4\frac{1}{2} \pm 1\frac{1}{2}\%$) is to reduce the compressive strength of the concrete but at the same time to improve the workability.

Air-entraining agents should not be confused with such compounds as aluminium powder which is used for the production of aerated lightweight concrete and mortar.

1.7.4 Plasticisers/Workability Aids
These admixtures can be divided into two main types:

Lignosulphonates, also known as lignins, and soaps or stearates
Finely divided powders

The lignosulphonates and stearates act very largely as lubricants and in this way the amount of water required in the mix to obtain a predetermined workability can be reduced; or, for a given w/c ratio, the workability is increased. Some of these compounds, namely the stearates, impart a degree of water repellency to the concrete, mortar or grout and these are sometimes referred to as waterproofers. An overdose of certain of these compounds will produce a set-retarding effect; serious overdosing can result in a permanent reduction in compressive strength.

The finely divided powders include pulverised fuel ash (PFA), powdered hydrated lime, powdered limestone and bentonite. Portland cement itself is a good plasticiser and an increase in cement content may help to overcome problems of segregation and harshness. Depending on the characteristics of the powder and quantity used, the water demand of the mix may be increased but its cohesiveness may be improved.

To sum up, both types (lignins and stearates, and powders) have their specific uses and both can be considered as reliable provided they are correctly used. They help to achieve a workable and cohesive mix which can be compacted under the action of poker vibrators and vibrating beams.

It should be remembered that many manufacturers of admixtures produced compounds which serve more than one purpose; this means that one can obtain plasticisers which also act as retarders, while other types act as accelerators. It is therefore important to obtain complete information on the basic composition and all the effects of a particular admixture before deciding to use it.

1.7.5 Superplasticisers
These materials are a relatively new type of chemical admixture as far as their use in the UK is concerned. They have, however, been in commercial use in Japan since about 1967 and in Germany since about 1972. It is important to appreciate that superplasticisers are distinct from the normal workability aids at present in general use; they can be used with confidence at high dosage levels, subject to the general conditions set out in this section.

Superplasticisers can be used for two purposes:

1. To produce a concrete having a virtually collapse slump, i.e. a 'flowable' concrete.
2. To produce a concrete with normal workability but with a very low w/c ratio, resulting in a high strength concrete.

It is not the intention in this section to discuss the chemical composition of the various types of superplasticisers available in the UK, but this can be summarised as follows:

Group 1—sulphonated melamine formaldehyde
Group 2—naphthalene sulphonated formaldehyde condensate
Group 3—modified ligno-sulphonates

Of these, the first two groups appear to be the more reliable and effective.

Research work and site tests in Germany, Japan and the UK have shown that so far the use of these compounds has not revealed any adverse effects on the durability of the concrete, nor on the ability of the concrete to passivate steel reinforcement and protect it from corrosion, nor has any reduction in the strength of the concrete over a long period of time been detected.

In order to obtain concrete which will 'flow' but not segregate, it is necessary to start with a slump of about 50–75 mm before the superplasticiser is added. The fine aggregate (sand) content of the mix must be increased by about 4–5% and the coarse aggregate correspondingly reduced so that the aggregate/cement ratio remains unchanged. The superplasticiser is added to the mix after the addition of the water and then mixing should be continued for at least another 2 min.

Correctly made superplasticised concrete has a slump of about 200 mm (when it is required to be flowable), and is almost self-levelling; it is cohesive and will not segregate. Strict site control of the mix proportions, especially the sand content and the original slump, are essential if segregation is to be avoided.

Because of the danger of segregation when using these admixtures, it is most important that trial mixes be carried out prior to their use and to ensure that the final mix design, including the type and grading of the fine and coarse aggregates, is exactly followed when the concrete is made.

It should be noted that maximum workability is usually only retained for a period of about 30–60 min, and then the concrete rapidly reverts to normal conditions of slump. The information so far available on superplasticisers shows that they are a very useful addition to the concreting industry. One of the basic principles of good quality concrete in the structure is that it must be thoroughly compacted. Full compaction can often prove difficult and in some cases almost impossible to achieve if high strength is to be maintained. This is particularly so when repairs with concrete have to be carried out to walls and columns as well as members of small section.

For further information on superplasticisers the reader is referred to the Bibliography at the end of this chapter.

1.7.6 Pulverised Fuel Ash (PFA)

PFA is produced in very large quantities from coal burning power stations. It is a very fine powder, having a specific surface similar to that of ordinary Portland cement, namely about $340 \, m^2/kg$. This is a new terminology for specific surface, which used to be expressed in cm^2/g, and for OPC, this would be $3400 \, cm^2/g$. The specific gravity of PFA is appreciably lower than cement, being in the range 1·9–2·3, while cement is about 3·12. The main compounds in PFA are oxides of silicon, iron and aluminium, together with some carbon and sulphur.

The relevant British Standard is BS 3892, and is in two Parts. Part 1:1982 relates to the use of PFA in structural concrete, and Part 2:1984 relates to its use in grouts and 'for miscellaneous uses in concrete'.

Part 1 contains tests and limits on:
Loss on ignition
Magnesia (magnesium oxide)
Sulphuric anhydride (SO_3)
Moisture content
Fineness
Water requirement of a mixture of PFA and OPC.

Part 2 contains similar tests, but with some differences in the pre-scribed limits, and the test for water requirement is omitted. Both Parts of the Standard include some general recommendations for the use of PFA.

In addition to the recommendations for use in the Standard the author considers the following are relevant:

(a) PFA should only be incorporated into concrete, grout or mortar with the written approval of the client's technical representative. The amount of PFA should be clearly stated.

(b) For concrete and mortar and grout, where the durability of steel reinforcement is concerned, the inclusion of PFA should not result in a reduction of the cement content below that required by the relevant specification.

(c) If PFA is used to replace some of the cement (subject to (b) above) it is likely that the striking times for formwork will have to be lengthened, and the period of curing increased compared with a similar mix without replacement.

For any engineer considering the use of PFA in concrete, mortar or grout, careful reading of the relevant Part of BS 3892 is strongly recommended.

It should be noted that the fineness of PFA is controlled and defined by sieving and not by reference to specific surface. Also, once PFA has been incorporated into a cement based mix, it is not possible to carry out any check on the amount added. Some experienced engineers consider it prudent not to allow the general use of PFA in cement based mixes.

1.7.7 Condensed Silica Fume

Condensed silica fume is a waste product of the ferrosilicon industry. It consists of 88–98% silicon dioxide with very small percentages (usually a maximum of 2% each) of carbon, ferric oxide, aluminium oxide (alumina) and oxides of sodium, potassium and magnesium.

It is an extremely fine greyish powder with a specific surface many times that of ordinary Portland cement. It is a highly reactive pozzolan, and a great deal of research and development in the laboratory and in the field has been carried out on the short and long term effects of using it as an admixture in concrete and mortar. A few references are included in the Bibliography at the end of this chapter but some brief notes are given below.

The addition of silica fume to concrete has a significant effect on the properties of the plastic concrete and imparts a number of beneficial characteristics to the hardened concrete and mortar. The dosage is likely to be in the range of 2–10% by weight of cement. Trial mixes are essential, and trial placements are recommended, as far as possible under site conditions. It is normally used with a lingosulphonate plasticiser or a superplasticiser. With experienced mix design techniques, very high strengths can be obtained (100 N/mm^2). In correctly designed mixes permeability is reduced and the highly reactive nature of the silica enables it to react with the calcium hydroxide in the hydrating cement paste, forming complex silicates which are more chemically resistant.

Reference to the Bibliography will show that there is reason to believe that the use of silica fume in concrete will be beneficial when there is a danger of alkali–silica reaction.

The use of silica fume in concrete and mortar in the UK has only been considered in recent years, and as is usual with new materials there is a tendency for exaggerated claims to be made for it. It is sometimes referred to as 'microsilica' and this expression has recently appeared as a trade name.

A very sophisticated version of silica fume, which contains among other things a high quality superplasticiser, is Corrocem, made in Norway by Norcem, the Norwegian cement manufacturers. It has been used effectively in the UK, USA, and on the Continent and in Scandinavia as an admixture in concrete which is in contact with ammonium-based fertilisers, and in dairies and sugar factories.

1.8 BONDING AIDS

Since the beginning of the 1960s in the UK and certainly earlier in the US, certain polymers, either on their own ór mixed with cement, have been increasingly used to improve bonding between hardened concrete and newly placed cement-based materials such as concrete and mortar. In the UK the three compounds in more general use are polyvinyl acetate (PVA), styrene–butadiene rubber (SBR) and acrylics. It is essential to obtain the best possible bond at the interface between existing concrete and mortar or concrete used for repair, and prior to the introduction of polymer bonding aids, it was the practice either to use nothing and rely on the preparation of the surface of the base concrete, or to use a cement slurry. Both of these techniques gave excellent results in the laboratory, but in the field, the results were often disappointing. This was the basic reason for the research and development into the use of special bonding aids.

It must be appreciated that the bond at the interface between the concrete and the repair material is likely to be subjected to considerable stress arising from changes in moisture content, freeze-thaw, a wide temperature range, as well as the force of gravity, and sometimes vibration.

Two tests for the effectiveness of the bond are the Slant Shear Test (BS 6319, Part 4, and ASTM C 882—78), and the pull-off test for which there is no UK Standard. The Slant Shear test can only be applied in the laboratory while the pull-off test can be applied both in the laboratory and on site.

The effect of weathering on the bond is obviously of great importance, and until laboratory specimens prepared for the Slant Shear test have been exposed to the weather for long periods, the effect of weathering can only be tested by pull-off tests on site. The author's experience is that when properly used, SBR/cement slurry helps to mitigate the inevitable effects of variation in workmanship and site conditions, and thus to improve the bond.

The vast majority of concrete repairs involve the placing of new concrete or mortar around cleaned reinforcement, and the high pH of SBR/cement slurry (about 12·0) helps considerably to ensure passivation of the rebars.

1.9 JOINT FILLERS AND SEALANTS

1.9.1 Fillers

Fillers for joints are sometimes known as 'backup' materials. They are used in full movement joints to provide a base for the sealant and also to prevent the ingress, during the construction period, of stones and debris which may prevent the joint from closing. Materials for these fillers include specially prepared fibres, cellular rubber and granulated cork compounds. The material used should fulfil the following requirements:

(a) It must be very durable.
(b) It must be chemically inert.
(c) When in contact with potable water it must be non-toxic and non-tainting and should not support bacterial or fungoid growth.
(d) It must be resilient and should not extrude so as to interfere with the sealing compound, and should not bond to the sealant as this latter could induce undesirable stress in the sealant.
(e) It should be easily formed to the correct dimensions and be readily inserted into the joint.
(f) In certain cases, e.g. floors, it should provide proper support for the sealant but should not bond to it.

Materials are now available which fulfil the above conditions reasonably satisfactorily.

1.9.2 Sealants
The materials used to seal joints in structures can be conveniently divided into two basic groups:

    (a) Preformed materials
    (b) Insitu compounds

To be satisfactory, both groups should possess the following characteristics:

    (i)   For external use, or in liquid retaining structures, the sealant itself must be impermeable.

    (ii) It must be very durable as periodic renewal may be difficult and expensive. Ideally, its life should be the same as the structure of which it forms a part; this condition is not fulfilled by any known sealant at the present time.

    (iii) It must bond to the sides of the groove in which it is inserted. In practical terms this means that the sealant should bond well to damp concrete.

    (iv) In potable water tanks the material must be non-toxic and non-tainting and should not support the growth of bacteria and fungi, etc.

    (v) As the joint opens and closes, the sealant must deform in response to that movement without undergoing any change which will adversely affect its integrity.

    (vi) It should be comparatively easy to install under the weather and site conditions relevant to the location of the structure.

The sealant, whether preformed or insitu, is normally accommodated in a groove in the concrete. Research in both the UK and the USA has shown that the shape and dimensions of the groove are important in ensuring a satisfactory and durable seal. For detailed information on this subject reference should be made to the Bibliography at the end of this chapter.

A preformed channel-shaped gasket made of EPDM has come onto the market. This spans the joint and is fixed with an adhesive into narrow grooves cut parallel to each side of the joint.

Preformed Materials
Preformed joint sealants at present take up only a small percentage of the market, but this share is steadily increasing. The majority of the high grade material is based on neoprene or EPDM and is imported from

the Continent and the USA. Volume for volume, neoprene or EPDM is cheaper than the high quality insitu sealants such as polysulphides and silicone rubber, but the cost of accurately forming the joint to receive the preformed strip reduces this margin.

When correctly installed, proprietary cellular neoprene or EPDM strips will remain watertight against pressures of up to 3 atm (30 m head of water). Indications, based on tests at Northampton sewage treatment works, showed that neoprene was the most durable of the sealants and is particularly resistant to attack by bacteria and mould growths.

Insitu Compounds

Insitu compounds are divided into a number of types, namely:

(a) Mastics
(b) Thermoplastics—hot applied
(c) Thermoplastics—cold applied
(d) Thermosetting compounds—chemically curing
(e) Thermosetting compounds—solvent release

The author is indebted to the American Concrete Institute for much of the information which follows.

a. Mastics. Mastics are generally compounds of a viscous liquid with the addition of fillers or fibres. They maintain their shape and stiffness by the formation of a skin on the surface and do not harden throughout the material nor set in the generally accepted use of the term. The vehicles, i.e. the viscous liquids, are usually low melting point asphalts, polybutylene, or a combination of these. They are used where the overriding factor is low first cost and where maintenance and replacement costs are not considered important. The extension—compression range is small and so these materials should only be used where small movement is anticipated.

b. Thermoplastics—hot applied. These materials become fluid on heating and on cooling they become an elastic solid, but the changes are physical only and no chemical reaction occurs. A typical example of this type of sealant is the rubber–bitumen compounds which are used extensively in many countries.

As the sealant has to be applied in a semi-liquid state it is only suitable for horizontal joints; it is used largely for roads and airfield pavements, but can also be used in the floors of reservoirs and sewage tanks.

The movement range which this type of material can accommodate is rather greater than that obtained with mastics, but is still small compared with thermosetting chemical curing elastomers.

There is a British Standard for 'Hot Applied Sealing Compounds for Concrete Pavements', BS 2499. The Standard is strictly a performance specification and says nothing about the chemical composition of the sealants. When used for tanks to hold potable water the sealant must be non-toxic and should not contain phenol compounds.

c. Thermoplastics—cold applied. These materials set and harden either by the evaporation of solvents (solvent release) or the breakup of emulsions on exposure to air. Sometimes a certain amount of heat is applied to assist workability, but generally they are used at ambient air temperature.

For water-retaining structures the most popular type is a rubber-asphalt. The movement which this type of sealant can accommodate is small; it hardens with age and suffers a corresponding reduction in elasticity. There is no British Standard for this type of sealant. A check should be made to ensure that the sealant does not contain toxic or phenol compounds and will not support bacterial or fungoid growth, before accepting its use in potable water tanks.

d. Thermosetting compounds—chemically curing. Materials in this category are one- or two-component compounds which cure (mature or harden) by chemical reaction to a solid state from the liquid or semi-liquid state in which they are applied.

High grade materials in this class are flexible and resilient and possess good weathering properties; they are also inert to a wide range of chemicals.

These compounds include polysulphides, polyurethanes, silicone rubber and epoxide based materials. They can be obtained so as to have an expansion–compression range of up to ±25% and a temperature range from −40°C to +80°C.

While they are considerably more expensive than the mastics and thermoplastics, they will accommodate far greater movement and are more durable.

In the UK the polysulphides are the most popular and can be used with confidence in a wide range of liquid-retaining structures, including those containing potable water. Some are used with a primer and some without. It is important to ascertain whether the particular brand selected will bond to damp concrete or whether a dry surface is required. Complete adhesion between the sealant and the sides (but not the base) of the sealing groove is essential for a liquid-tight joint. In the climatic conditions of the UK it is virtually impossible to ensure dry concrete on most construction sites. Two-part polysulphide-based sealants are covered by BS 4254, which deals with two grades, a pouring grade and a gun grade. One-part gun grade polysulphide-based sealants are covered by BS 5215.

e. Thermosetting compounds—solvent release. Sealants of this type cure by the release of solvents present in the compound itself. The principal materials in use are based on such compounds as butyl, neoprene and polyethylene. Their general characteristics are somewhat similar to those of solvent-release thermoplastics; their extension–compression range is about ±7%.

There is no British Standard for this type of sealant.

1.10 ORGANIC POLYMERS

Organic polymers are complex chemical compounds derived mainly from the petrochemical industry. These materials are often referred to as 'resins', and the principal resins used in the construction industry are epoxide, polyurethane, polyester, acrylic, polyvinyl acetate, and styrene–butadiene. The basic raw materials are supplied by comparatively few manufacturers, such as Shell Chemicals, CIBA, Dunlop Chemical Products Division, Revertex, Borden Chemical Co. and Dow Chemicals. The raw materials are then taken by a large number of formulators who formulate the final products in such a way that they possess specific characteristics needed for the use to which they are put.

Under the particular condition of curing, which in some cases requires hardeners or accelerators, the resins form long molecular chains in three dimensions, which can result in an extremely strong and stable material.

While the range of use of these materials is now very wide, it is convenient and practical to divide it into two main categories; namely, coatings in which the formulated compound is used on its own, and mortars and concretes in which the resin is mixed with aggregate and sometimes cement.

It is quite usual to find that several types of polymer are used in combination in order to obtain the optimum results. The information given here is intended for practising engineers and nor for chemists.

For coatings used for the protection of concrete and to reduce permeability, epoxies, polyurethanes and acrylics are in most general use. For use in mortars and concretes, acrylics and styrene–butadiene compounds are used successfully. Polyvinyl acetate (PVA) is used as a bonding agent for floor screeds and toppings to increase adhesion with the base concrete; it is also used in cement mortar mixes to improve certain characteristics of the mortar and bond with the substrate. Polyester resins are used in cement-based proprietary floor toppings.

1.10.1 Epoxide Resins

The resins are marketed by the formulators and have the special properties required for the specific use to which they will be put. For example, some resins can be successfully applied and cured under water. While most epoxies are rigid when cured, it is now possible to obtain a type which is slightly flexible.

The basic characteristics of epoxide resins include the following:

(a) Outstanding adhesive qualities to such materials as concrete and steel.

(b) Resistance to a wide range of acids and alkalis and other chemicals, except acids such as nitric which have high oxidising characteristics.

(c) Rather vulnerable to organic solvents.

(d) Low shrinkage when the compound cures and changes from the liquid to the solid state.

(e) High coefficient of thermal movement compared with concrete. The coefficient of thermal expansion of the pure resin is very high, about 60×10^{-6}; when used as a mortar, the mixture has a coefficient of about $25-30 \times 10^{-6}$. When used as a coating, the coefficient depends mainly on the amount of filler used. A Portland cement/sand mortar has a coefficient in the range $7-12 \times 10^{-6}$.

(f) High compressive, tensile and flexural strength.

(g) Appreciable loss of strength at temperatures over about 80°C.

(h) High rate of gain of strength, which can be varied to suit the particular application.

(i) Rather poor resistance to fire compared with concrete and clay bricks.

(j) To obtain satisfactory results, the site conditions under which the resin will be applied must be stated to the formulator. Unless specially formulated, most epoxide resins must be applied to dry surfaces and the ambient air temperature and relative humidity during application must be kept within fairly narrow limits.

The resins are usually two-pack materials, consisting of a basic resin and an accelerator (sometimes called a hardener or activator). The two materials must be thoroughly mixed immediately prior to use.

Some of the more important factors relating to the use of epoxides are:

1. *Pot life*. This is the period which can be allowed to elapse between the mixing of the resin and the activator and the completion of the application of that particular mix. This period can be varied by the formulator to suit site conditions, but increasing the pot life will result in a slowing down of the rate of hardening of the applied coating. The period usually adopted is fairly short, namely a few hours, but the actual range can be between 30 min and about 48 h.

2. *Hardening*. This is the physical setting of the plastic resin after application; it can be varied by the formulator. It is generally recommended that each coat must harden before the next coat is applied, and therefore applicators must acquaint themselves with the hardening characteristics of the particular resin they propose to use.

3. *Curing*. Curing is the term used to describe the maturing and gain of strength of the resin. It is really the formation of the molecular linkage which imparts the desirable strength and durability to the final product. The curing period can also be varied by the formulator, but an 'average' period is about seven days. Curing usually ceases when the ambient air temperature falls to about 5°C and this can cause some difficulties on site when these resins are used and early strength and bond are required.

There are about fifteen different types of epoxide resins on the market and probably about three hundred hardeners, so that the possible number of combinations of resin and hardener is very high. For this reason it is inadvisable for anyone who is not an experienced polymer chemist to attempt to advise on a specification for an epoxide resin for a particular purpose. The basic requirements for the use of the resin must of course be clearly laid down, such as setting time, curing time, ability to bond to wet concrete, etc.; the details of how these requirements are met must be left to the formulator.

1.10.2 Polyurethanes

Polyurethanes, like epoxide resins, are products of the petrochemical industry. They can be obtained as elastomers, solid and rigid materials and as flexible coatings. They are very durable in external conditions and retain their gloss well. For use in the construction industry they are usually formulated as two-pack and single-pack materials. Generally, the two-pack material has better durability than the single-pack. Polyurethanes can be specially formulated to meet specific site requirements. One characteristic of particular importance is that they will cure in temperatures appreciably below 0°C, whereas epoxide resins cannot be relied upon to continue to gain in strength (cure) when the temperature falls to 5°C and below. They can be combined with epoxies and will withstand relatively high temperatures as well as sudden changes in temperature, i.e. thermal shock.

1.10.3 Polyester Resins

Polyester resins are used with Portland cement and selected aggregates to form a polymer–cement–aggregate mortar. Such mortars possess a number of desirable qualities such as good resistance to a wide range of chemicals, high resistance to abrasion, waterproofness under the range of heads likely to be met in liquid-retaining structures, and high bond strength with most common building materials.

They are also used with glass fibre to form linings to various types of liquid-retaining structures and liquid-conveying systems such as large diameter sewers.

There are a number of important differences between the properties of polyester resins and epoxide resins. The polyesters can be used in a wider temperature range; they offer rather better resistance to heat, but have appreciably higher shrinkage characteristics; the bonding properties to concrete are generally lower. By adjusting the ratio of resin to catalyst, the hardening time can be made very short, and once a 'set' occurs the rate of gain of strength is very rapid. The coefficient of thermal expansion for a polyester mortar would be in the range $25–35 \times 10^{-6}$, compared with $7–12 \times 10^{-6}$ for a Portland cement/sand mortar.

1.10.4 Polyvinyl Acetate (PVA)

This material is used as a bonding agent and as an admixture in mortar to improve certain properties of the mortar. Manufacturers of proprietary compounds based on PVA claim increased tensile and flexural strength, reduced drying shrinkage and reduced permeability. The author considers that some of the improvements claimed are marginal and the behaviour of PVA under permanently damp conditions can be very disappointing.

1.10.5 Styrene–Butadiene and Acrylic Resins

These materials are also known as latexes and polymer emulsions. Some of the properties of a proprietary styrene–butadiene copolymer emulsion (latex) used with Portland cement in grout, mortar and concrete are:

pH	11·0
Total solids content	47% (nominal)
Specific gravity	1·1
Butadiene content	40% (nominal)

The author is indebted to Doverstrand Ltd for the above information. Acrylic latexes (styrene acrylic and all acrylic) generally have shorter setting and hardening times and greater resistance to ultraviolet light than the styrene–butadiene type, and are used more on the Continent than in the UK.

The advantages claimed for the use of these latexes in Portland cement grout, mortar and concrete include reduced permeability, reduced initial drying shrinkage, improved resistance to attack by certain dilute acids and solutions of sulphates, and improved bond to the substrate.

Special precautions have to be taken by formulators with styrene–butadiene latexes when these are used with OPC to reduce air entrainment and the effect of the retardation of the cement.

1.11 POLYMERISED CONCRETE

This is sometimes rather loosely referred to as polymer concrete. In this book polymerised concrete means Portland cement concrete containing a monomer and which is polymerised after it has hardened. On the other hand polymer concrete is concrete in which the cement is replaced either entirely or principally by an organic polymer such as epoxide or polyester resin or normal concrete which contains polymer as an admixture.

Polymerised concrete can be divided into two types:

(1) The complete unit of hardened concrete is impregnated (usually by dipping) with a monomer and is then polymerised either by heat or gamma rays.

(2) The monomer is mixed with the gauging water and then after the concrete unit has hardened it is polymerised by heat.

A considerable amount of work has been carried on for some years in a number of countries into the techniques of polymerising concrete and the characteristics of the finished concrete. Claims made for polymerised concrete include the following:

(a) A considerable increase (up to four times) in the compressive and tensile strength.
(b) The resistance to chemical attack and the effects of freeze-thaw are greatly increased.
(c) Absorption and permeability are greatly reduced.

All the above increases and decreases are related to control specimens of normal Portland cement concrete.

The process of producing polymerised concrete is complicated and expensive and so far it appears to have been used only on a small scale in selected locations for long term test purposes. Some references on this material are given in the Bibliography at the end of this chapter.

1.12 GLASS FIBRE REINFORCED PLASTICS

This material is usually referred to as GRP. It is a composite material composed of polyester resin and glass fibre. There are various types of glass fibre and various types of polyester resin. The manufacturers select the most suitable type for a particular application. The resin and the glass fibre are the essential materials and form the bulk of the final product but there are in addition fillers, pigments, stabilisers and numerous other additives to impart special properties to finished material.

The author has not seen any published information on the use of this material for the repair of concrete, but he has seen examples of GRP linings applied to liquid retaining structures to remedy leakage and to protect the concrete against chemical attack.

The process of application consists in building up successive layers of the resin and glass fibre on the substrate. This can be done by hand or spray, but the latter is now more usual. With spray application, the resin, catalyst and glass fibre are applied through a three-nozzle gun. It is important that the glass fibres are completely covered by the resin, and special techniques are employed to ensure this. Normal good practice for the preparation of the concrete surface prior to application of the resin/glass fibre matrix is essential.

Some detailed information on GRP linings to concrete swimming pools (generally installed to remedy leakage) is given in the author's book, *Swimming Pools*.

Points to remember when using this material for repairs to concrete are:

(a) Polyester resins have high shrinkage characteristics, and the coefficient of thermal expansion of the resin is significantly different to that of the glass fibres.

(b) It can be difficult to obtain a permanently good bond with the concrete substrate.

(c) If cracks occur in the GRP, moisture penetration can result in delamination and debonding.

1.13 FLEXIBLE SHEETING

Various types of sheeting can be used with advantage in waterproofing concrete structures. The better known materials which are in general use are:

Butyl rubber
Chlorinated polyethylene (CPE)
Polyvinyl chloride (PVC)
Polyethylene (sold under various trade names)
Chlorosulphonated polyethylene (trade name: Hypalon).

All the above materials have advantages and disadvantages and the specifier should decide on the basic requirements, which may include all or some of the following:

Size of sheets and range of thicknesses
Method of making joints
Method of fixing
Types of adhesives required
Aging under anticipated conditions of exposure
Resistance to abrasion
Flexibility (extension and recovery)
Resistance to chemical attack (when applicable).

BIBLIOGRAPHY

AMERICAN CONCRETE INSTITUTE, *Guide to Joint Sealants for Concrete Structures*, ACI Committee 504, ref. 504R–77, 1977, p. 58.

AMERICAN CONCRETE INSTITUTE, *Cement and concrete terminology*, ACI Committee 116, ref. SP–19(78), 1978, p. 50.

AMERICAN CONCRETE INSTITUTE, ACI Committee 503, Four epoxy Standards, 503.1–79 to 503.4–79, 1979, p. 24.

AMERICAN CONCRETE INSTITUTE, *Cedric Wilson Symposium on Expansive Cement*, ref. SP–64, 1980, p. 336.

AMERICAN CONCRETE INSTITUTE, *Use of epoxy compounds with concrete*, ACI Committee 503, ref. 503R–90, 1980, p. 33.

AMERICAN CONCRETE INSTITUTE, *Applications of polymer concrete*, ref. SP–69, 1981, p. 228.

AMERICAN CONCRETE INSTITUTE, *Superplasticisers in concrete*, collection of 30 papers from 11 countries, ref. SP–68, 1981, p. 572.

AMERICAN CONCRETE INSTITUTE, *State-of-the-Art Report on fiber reinforced concrete*, ACI Committee 544, ref. 544 1R–82, 1982, p. 16.

AMERICAN CONCRETE INSTITUTE, *Guide for the use of admixtures in concrete*, ACI Committee 212, ref. 212.2R–81, p. 42.

ASGEIRSSON, H. AND GUDMUNDSSON, G., Pozzolanic activity of silica dust *Cement & Concrete Res.*, **9**(2), March 1979, 249–52.

BARNES, P. (Ed.), *The Structure Performance of Cements*, Applied Science Publishers Ltd, London, 1983, p. 560.

BATES, S. C. C., *High Alumina Cement Concrete in Existing Building Superstructures*, HMSO, London, 1984, p. 109.

BEECH, J. C., *The Selection and Performance of Sealants*, Building Research Establishment, Information Paper IP.25/81, December 1981, p. 4.

BRITISH STANDARDS INSTITUTION, Code of Practice CP 3003, *Lining of vessels and equipment for chemical processes*, Parts 1, 4, 5 and 6.

BRITISH STANDARDS INSTITUTION, *Glossary of terms used in concrete and reinforced concrete*, BS 2787.

BRITISH STANDARDS INSTITUTION, *Glossary of building and civil engineering terms*, BS 6100, Parts 1–6.

BRITISH STANDARDS INSTITUTION, *Glossary of terms used in non-destructive testing*, BS 3683.

BRITISH STANDARDS INSTITUTION, *Glossary of terms used in plastering, rendering and screeding*, BS 4049.

BRITISH STANDARDS INSTITUTION, BS 4550, Parts 1 to 6, *Methods of testing cement*.

BRITISH STANDARDS INSTITUTION, *Code of Practice for the structural use of concrete*, BS 8110, Parts 1 and 2.

CEMBUREAU, *Cement Standards of the World, Portland Cement and its Derivatives*, Cembureau, Paris, 1985.

CEMENT ADMIXTURES ASSOCIATION, Admixture Data Sheet, published by the Association, Southampton, March 1982.

CEMENT & CONCRETE ASSOCIATION, *Superplasticizing Admixtures for Concrete*, published by the Association, 1976, p. 31.

CONCRETE SOCIETY, *Guide to Chemical Admixtures*, Technical Report No. 18, published by the Society, 1980, p. 16.

CONCRETE SOCIETY, *Changes in cement properties and their effects on concrete*, Report of a Working Party, Sept. 1984, p. 22.

COOK, D. J. AND CROOKHAM, G. D., Fracture toughness measurements of polymer concretes. *Mag. Conc. Res.*, **30**(105), Dec. 1978, 205–14.

DENNIS, R., Latex in the construction industry, *Chemistry & Industry*, Aug. 1985, 505–12.

INTERNATIONAL LEAD ZINC RESEARCH ORGANIZATION INC., *Galvanized Reinforcement for Concrete*, Vol. 2, New York, May 1981, p. 208.

KAYS, W. B., *Construction of Linings for Reservoirs, tanks and Pollution Control Facilities*. John Wiley, New York and London.

LEA, F. M., *The Chemistry of Cement and Concrete*, 3rd ed., Edward Arnold Ltd, London, 1970, p. 725.

MALHOTRA, V. M. AND CARETTE, G. C., *Silica fume concrete, properties, applications and limitations*, Institute of Concrete Technology, 10th Annual Convention, Slough, June 1982, Paper No. 2, p. 22.

MIDGLEY, H. G. AND MIDGLEY, A., The conversion of high alumina cement. *Mag. Conc. Res.*, **27**(91), June 1975, 59–77.

NEVILLE, A. M., *Properties of Concrete*, 3rd ed., Pitman & Sons Ltd, London, 1981, p. 779.

PERKINS, P. H., *Swimming Pools*. Applied Science Publishers, London, 1971, p. 358.

PERKINS, P. H., The use of SBR/cement slurry for bonding coats, *J. Conc.*, **18**(3), March 1984, 18, 19.

Proceedings, of First International Congress on Polymer Concrete, London, May 1975; Concrete Society, ACI, RILEM, British Plastics Federation; Construction Press, London, and Concrete Soc., London.

PRODORITE LIMITED, *Corrosion Resisting Cements*, published by Prodorite Ltd, Wednesbury, England, 1984.

Sealant Manufacturers' Conference, *Manual of Good Practice in Sealant Application*, Jan. 1976, published jointly by Sealant Manufacturers' Conference and CIRIA, London, p. 72.

TABOR, L. J., *The effective use of epoxy and polyester resins in civil engineering structures*. CIRIA Report no. 69, 1978, p. 68.

WOLSIEFER, J., *Ultra-high strength field placeable concrete*, Paper at ACI Annual Convention, Atlanta, Jan. 1982, p. 23.

ZINC DEVELOPMENT ASSOCIATION, *Galvanizing Guide*, London, 1979, p. 58.

Chapter 2

Factors Controlling the Deterioration of Concrete

It is intended in this chapter to consider the more important factors which can cause deterioration in reinforced concrete structures. It is only by carefully reviewing why and how this deterioration occurs that satisfactory techniques can be developed for the repair. This review will also be useful in helping to prevent this deterioration from taking place in future structures. The accumulation of knowledge and experience is a continuous process. Mistakes are inevitable; they should also be forgivable provided the necessary lessons are learned.

At the present time there is no material known which is completely inert to chemical action and immune to physical deterioration. While concrete is no exception to this, it has, under what may be termed 'normal conditions of exposure', a very long life. Concrete made from naturally occurring cements (pozzolanic materials such as trass) has been found in excellent condition after more than 2000 years. There is no reason to believe that under similar conditions, modern Portland cement concrete will have a shorter life. This should satisfy even the most conservative client! However, it is known that present day environmental conditions can be aggressive to concrete and other materials used in construction.

Only two basic types of material will be considered in this chapter, namely the concrete itself and metals which are usually used in conjunction with it. The metals most commonly used are various grades of steel (ferrous metals including galvanised steel and stainless steel), aluminium, copper, phosphor-bronze and gunmetal.

In essence, deterioration can be due to chemical attack on or between the various materials of which the structure is built, or physical deterioration arising from climatic changes, abrasion, damage from high velocity water, fire, impact, explosion, foundation failure or overloading.

It is essential to understand how these various causes are likely to reveal themselves, so that a realistic assessment of the overall cause(s) can be made and a proper diagnosis arrived at.

2.1 DURABILITY AND PERMEABILITY OF CONCRETE

It is the author's experience that a very high percentage of the defects in concrete structures arise from the corrosion of the reinforcement. Therefore in the design and construction of new structures and in the repair of existing ones, special attention must be given to ensuring the long term protection of the reinforcement. Unless stainless steel is used, or the rebars are given a complete, durable and impermeable coating such as heavy galvanising or special powder epoxy (see Chapter 1), the concrete surrounding the rebars must be as impermeable as possible. There are sometimes misunderstandings on the practical meaning of the terms 'durability' and 'permeability' and therefore some observations are given below to define the words used for the purpose of this book.

2.1.1 Durability
A structure would be considered durable if it fulfilled its intended duty for the whole of its design life with the minimum of maintenance. The design life of the structure will usually be laid down by the client in consultation with the designer. It would be unrealistic to expect any structure to maintain its 'as new' condition without any maintenance whatever.

However, Portland cement concrete has the potential of an almost unlimited life unless it is subjected to chemical attack by an aggressive environment, or suffers physical damage. Weather staining and similar discoloration should not be confused with lack of durability. On the other hand, deep carbonation, chemical attack on the concrete, cracking and spalling due to poor quality materials or workmanship, and/or corrosion of the reinforcement, would be a clear case of low durability.

Concrete structures never consist entirely of concrete, as all these structures contain other materials which vary with the type of structure and include reinforcement, joint sealants and fillers, thermal and sound insulation, metal fixings, pipe connections, and waterproofing and decorative layers. Some of these 'other materials' will have a limited life compared with the concrete. In particular, joint sealants, fillers, water-proofing and decorative layers will require periodic renewal.

The question of how to achieve durability in concrete is discussed in the next Section, in which it will be seen that durability is closely related to permeability. In some cases, certain parts of a structure may be subject to physical deterioration such as abrasion caused by steel wheeled trolleys on a floor, a jet or stream of high velocity water containing grit impinging on a concrete wall or floor, spalling and surface flaking due to freeze-thaw cycles, and damage by wave action abrasion by sand and shingle in the case of marine structures.

2.1.2 Permeability

For durability, it is accepted that concrete should possess low permeability. Unfortunately it has been found impossible so far to set limits for permeability which can be subjected to practical tests.

Permeability should not be confused with absorption. The permeability of concrete is not a simple function of its porosity but depends on the size, distribution and continuity of the pores. The size of capillary pores in concrete is about $1.3\,\mu\mathrm{m}$, and the gel pores are very much smaller. The volume of pore space in concrete, as distinct from its permeability, is measured by absorption, and the two quantities are not necessarily related.

Permeability tests measure the rate at which a gas or liquid passes right through the test specimen under an applied head. Concrete possesses a pore structure and in this respect is different to metals. The capillary pore structure allows water under pressure to pass slowly through the concrete, but the rate of flow through dense, good quality concrete is extremely slow. According to Neville's *Properties of Concrete*, cement gel has a porosity of 28%, but its permeability is about 7×10^{-16} m/s. The permeability of the cement paste as a whole is 20–100 times greater than that of the gel itself.

The subject of permeability is very complex, but most concrete engineers agree that the following are the main factors involved:

(a) The quality of the cement and aggregate.
(b) The quality and quantity of the cement paste; the quality of the cement paste depends on the amount of cement in the mix, the water/cement ratio, and the degree of hydration of the cement.
(c) The bond developed between the paste and the aggregate.
(d) The degree of compaction of the concrete.

(e) The presence or absence of cracking.

(f) The standard of curing.

(g) The characteristics of any admixtures used in the mix.

2.1.3 Carbonation

Carbonation is the effect of carbon dioxide (CO_2) in the air on Portland cement products, mainly calcium hydroxide ($Ca(OH)_2$) in the presence of moisture. The $Ca(OH)_2$ is converted to calcium carbonate ($CaCO_3$) by absorption of carbon dioxide. The calcium carbonate is only slightly soluble in water and therefore, when it is formed it tends to seal the surface pores of the concrete, provided the concrete is reasonably dense and impermeable.

The pH of the pore water in concrete is generally between 12·5 and 13·5 but if, due to carbonation, it is lowered to 9·0 and below, corrosion of the reinforcement may occur. Therefore the depth of carbonation in reinforced concrete is an important factor in the protection of the reinforcement; the deeper the carbonation, the greater the risk of corrosion of the steel. The extent of carbonation can be determined by treatment with phenolphthalein; the presence of alkalinity shows as a pink colour, while the carbonated part of the concrete remains without colour change.

Good quality dense concrete carbonates very slowly; even after a period of 50 years carbonation is unlikely to penetrate to a greater depth than 5–10 mm. On the other hand, a low strength, permeable concrete may carbonate to a depth of 25 mm in less than 10 years. Experience suggests that low quality cast stone products are particularly prone to carbonation. Carbonation does not adversely affect the durability of the concrete itself; it is the indirect effect it has on steel rebars that makes it undesirable in a reinforced concrete structure. There is reason to believe that carbonation of concrete tends to reduce permeability to the passage of moisture.

In recent years it has become the practice also to include in the term 'carbonation' the reaction between oxides of sulphur (sulphur dioxide and sulphur trioxide) and the calcium hydroxide in the cement. These oxides in solution in rain water and atmospheric moisture are acidic and therefore react with the alkalis in the cement paste in the same way as does carbon dioxide.

The author emphasises that in cases of deterioration of concrete, it is seldom that the concrete itself suffers significant chemical attack. It is not unusual to find that corrosion of steel reinforcement in a marine structure is incorrectly diagnosed as attack by the sea water on the concrete.

2.2 CHEMICAL AGGRESSION TO CONCRETE

2.2.1 General Considerations
Chemical attack on the concrete is likely to occur from one or more of the following causes:

(a) Aggressive compounds in solution in the sub-soil and/or ground water.
(b) Aggressive chemicals in the air surrounding the structure.
(c) Aggressive chemicals or liquid stored in, or in contact with, the structure.
(d) Chemical reaction between the constituents of the concrete, i.e., alkali–aggregate reaction; this is a special case.

The protection of concrete structures from chemical attack is dealt with in Chapters 6 and 9. Some brief explanations will be given here of the principles involved in the various types of chemical attack summarised above. From a practical point of view (and this book is intended for engineers and not chemists), the chemicals which attack concrete can be divided into four main categories:

(a) Acids (all).
(b) Ammonium compounds (some, not all).
(c) Sulphates (all).
(d) Other.

For attack to take place, the compounds must be in solution; for all practical purposes, there would be no chemical reaction between a piece of completely dry concrete and a piece of completely dry ammonium nitrate. However in practice, these 'completely dry' conditions do not exist; moisture is always present.

Another point of practical importance is that the chemical attack takes place on the cement rather than the aggregate except when the aggregate is calcareous and the attack comes from an acid.

2.2.2 Acids
The principle involved is that all acids react with alkalis, and Portland cement is highly alkaline. Similarly, acids will attack calcareous ag-

FIG. 2.1. (a) Concrete sewer pipes attacked by a trade effluent containing hydrochloric acid and other aggressive chemicals. Courtesy of Warrington & Runcorn Development Corporation.

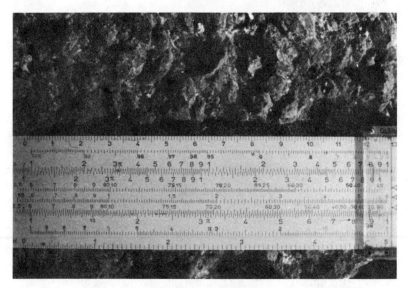

FIG. 2.1. (b) Concrete damaged by sulphuric acid.

gregates such as limestone. A solution is said to be acidic when the pH is below the neutral point of 7·0 and alkaline when the pH is above 7·0. The pH is the hydrogen ion concentration and the pH scale is logarithmic. This means that the hydrogen ion concentration of a liquid with a pH of 2·0 is 10 000 times that of the liquid with a pH of 6·0. It is important to realise that the pH alone does not define the type nor the amount of acid present, it measures the intensity of the acidity.

It is possible to calculate, theoretically, the approximate amount of cement neutralised by a solution of a given acid which has a stated pH value. In this way, the theoretical increase in the destructive power of two solutions of the same acid, as the pH value falls, can be demonstrated.

If the acid penetrates the concrete through cracks down to the rebars, then the steel will corrode and spall the concrete. The deterioration of the concrete would then be largely due to the corrosion of the steel rather than direct attack on the concrete.

An important point to remember is that the same concentration in solution of different acids gives different pH values. Figure 2.1(a) shows concrete sewer pipes attacked by a trade effluent containing hydrochloric acid and other aggressive chemicals. Figure 2.1(b) shows concrete severely damaged by sulphuric acid.

2.2.3 Ammonium Compounds

Not all ammonium compounds are harmful (for example, ammonium carbonate), but most are, particularly those used in chemical fertilisers such as ammonium nitrate, ammonium sulphate and ammonium superphosphate. The attack can be very severe, especially in warm, humid conditions.

While some authorities consider that reinforcement is at risk in concrete attacked by ammonium compounds, the writer has seen examples of concrete which has suffered severe attack by ammonium nitrate but with no corrosion of the rebars. The American Concrete Institute Committee 515 reports that while ammonia liquid is only harmful to concrete if it contains ammonium salts, ammonia vapours are likely to attack moist concrete slowly.

Lea, in the *Magazine of Concrete Research*, has reported that 5% solutions of ammonium chloride and ammonium nitrate attack Portland cement concrete by acting as dilute acids due to loss of ammonia by reaction with lime. The action of ammonium sulphate solution results in disintegration, which in this case is due mainly to the expansion caused by the formation of calcium sulpho-aluminate. High alumina cement concrete is much less affected by solutions of ammonium salts than Portland cement concrete. Further information on the effect of ammonium based fertilisers on concrete and on suitable protective measures is given in Chapter 6.

2.2.4 Sulphates

In theory, all solutions of sulphates will attack Portland cement concrete to a greater or lesser extent. The sulphate reacts with the tricalcium aluminate (C_3A) in the cement to form the compound ettringite, the reaction being expansive in character. The degree of attack depends on a number of factors, the principal ones being:

(a) The percentage of C_3A in the cement.
(b) The permeability of the concrete.
(c) The solubility of the sulphate in question.
(d) Whether the cations in the sulphate react with the compounds in the cement, as for example magnesium sulphate and ammonium sulphate; this constitutes double decomposition.

In the vast majority of cases, the sulphates are external to the concrete, i.e. in solution in ground water or trade effluent. However, in some parts of the world, e.g. the Middle East, sulphates are present in significant concentrations in the aggregates used for the concrete (generally the fine aggregate). The result is the same: the formation of ettringite and the expansive disintegration of the concrete (see Fig. 2.2).

If saline (brackish) water is used for mixing concrete, then the sulphate concentration in the mixing water must also be taken into account, and then the total sulphate content in the mix becomes the controlling factor. This is discussed in Section 2.5. It must be remembered that Portland cement itself contains sulphate as gypsum (calcium sulphate) which is used in the manufacture to control the set of the cement, but the amount is normally limited to about 3·0% by weight of cement. The total 'acceptable' concentration of sulphate in concrete is about 4·0% by weight of cement.

2.2.5 Other Aggressive Compounds

Due to the vast number of chemical compounds which exist in nature, as well as the products of the chemical industry, it is not possible to list them all nor to state whether or not they are likely to be aggressive to Portland cement concrete. A comprehensive list is given in the Report of

FIG. 2.2. Concrete damaged by severe sulphate attack.

Committee 515 of the American Concrete Institute, of November 1979. The author therefore proposes only to comment briefly on those compounds, in addition to those already mentioned, which in his experience are most likely to be met when dealing with problems of chemical attack on concrete.

Brine. This is usually a concentrated solution of sodium chloride (common salt). At normal temperatures, this is not aggressive to good quality concrete. Strong calcium chloride brine may cause some disintegration in the long term. Solutions of chlorides will attack steel severely. Also, if the salt solution enters the pores of the concrete and then crystallises out, some physical disruption of the concrete is likely to occur.

Calcium hypochlorite. A concentrated solution may attack concrete slowly, particularly if the concrete is rather porous. The solution is alkaline in reaction, and dilute solutions are not harmful to good quality concrete.

Caustic soda (*sodium hydroxide*). This and other caustic alkalis in concentrations up to about 10%, are not harmful. Higher concentrations, particularly if accompanied by a rise in temperature, can cause slow disintegration of concrete.

Distilled and demineralised water. As this is very pure water it is at first sight surprising that it would cause deterioration of concrete. However, it has very high dissolving power, and the main characteristics are:

Very low calcium hardness (virtually zero).
Very low total dissolved solids (virtually zero).
Low alkalinity.

The water may also contain dissolved carbon dioxide, and this would increase considerably its potential for attack on concrete. This type of water has a negative Langelier Index which means that it is 'lime

FIG. 2.3. Concrete severely damaged by hot distilled water (courtesy: Dr D. Higgins).

dissolving'; if the Index were positive, it would indicate that the water was 'lime depositing.'

It is advisable to assume that distilled and demineralised water will attack concrete. A more detailed, but brief discussion on waters with a negative Langelier Index is given in Chapter 9, Section 9.4.3. Figure 2.3 shows disintegration of concrete by hot distilled water.

Ferrous sulphate. This is acidic in reaction and will therefore attack concrete. The compound is sometimes used in contact tanks in water treatment works, and in dilute solution will etch the surface of concrete.

Fruit and vegetable juices. These contain organic acids and sugars, and therefore are aggressive to concrete.

Hydrogen sulphide. This is a gas and is also known as sulphuretted hydrogen. It is generated by anaerobic bacteria in septic sewage. The gas itself will not attack concrete, but it is readily soluble in water, and is oxidised to sulphurous and sulphuric acids which are very aggressive to concrete.

Milk. Fresh milk is not aggressive to concrete, but sour milk contains lactic acid which attacks concrete; the degree of attack depends on the permeability of the concrete and the concentration of lactic acid present.

Petroleum oils. These are generally not aggressive to concrete provided the acid and sulphur contents are low. They have high penetrating characteristics.

Sea water. Generally this is not aggressive to dense, good quality concrete. The concentrations of dissolved salts, mainly chlorides and sulphates, vary. In Atlantic water, are about 18 000 ppm of chlorides and about 2500 ppm of sulphates. In the Red Sea and Persian Gulf, the concentrations are at least 25% higher.

Sewage. Ordinary domestic sewage is not aggressive to concrete but trade wastes can attack concrete if they contain aggressive chemicals in solution, particularly acids. The severity of the attack depends on the concentration (dilution factor), period of contact with the concrete, and to some extent on the temperature.

Sugar. This will attack concrete slowly, but the experience of the author is that once the sugar solution penetrates into the concrete the rate of attack increases. Traffic on the floor, as in factories and warehouses, aggravates the situation.

Urea. There are differences of opinion as to whether the deterioration of concrete in contact with urea is due to chemical attack or whether it is physical, arising from the growth of crystals in the pores of the concrete. The general concensus of informed opinion is that the attack is not due to chemical reaction between the urea and the cement. Urea is used in the fertiliser industry and as a de-icing salt on roads and airfield runways. The chemical formula for urea is $CO(NH)_2$.

Sodium chloride is also used, in large quantities, as a de-icing salt, but is known to be very aggressive to steel reinforcement. From time to time, suggestions are made for the use of urea to replace sodium chloride. However, work in US by Forbes, Stewart and Spellman showed that urea is almost as aggressive to steel as sodium chloride.

A paper by A. Rauen, at the June 1980 International Colloquium on the Frost Resistance of Concrete, held in Vienna, contains clear statements concerning the action of urea which can result in significant deterioration of the concrete, quite apart from its corrosive effect on steel. From the Paper it appears that urea reduces the surface tension of water and so aggregates can be damaged by water penetrating into the pores and the urea crystallising out.

While it must be admitted that the case against the use of urea as a substitute for sodium and calcium chloride has not been completely worked out, the author feels that there is sufficient evidence to justify great caution.

2.2.6 Alkali–Aggregate Reaction (AAR)
It is believed that this type of deterioration in concrete was first identified in the USA in about 1940. Since then it has been found in many countries including Iceland, Denmark, Germany, Canada, Cyprus, Turkey, Australia and South Africa, and since 1976 on the mainland in the United Kingdom.

Since it was first diagnosed in the UK in concrete in some CEGB substations, and a multi-storey car park in Plymouth, the number of confirmed cases has increased year by year. It is difficult to obtain reliable figures, but the *New Civil Engineer* in an article in the issue of 2

May 1985 stated that official estimates for AAR in road bridges put the number of suspected cases at over 350. The Cement & Concrete Association in October 1985 put the number of confirmed cases of ASR at 60.

There are two forms of AAR, alkali–silica reaction (ASR) and alkali–carbonate reaction. The former (ASR) is much more common the world over, and the only one found so far in the UK. Both types result from the interaction of alkalis in the concrete (mainly originating in Portland cement) and certain types of siliceous aggregate.

The key factor is the alkalinity of the pore fluid in the concrete. This originates mainly but not exclusively from alkali metal salts in Portland cement. External sources of these metallic salts, e.g. sea water, and de-icing salts will increase the total alkalinity if they penetrate into the concrete but at present they are considered to be of minor importance. Irrespective of the amount of alkali in the concrete, the reaction (AAR) will only occur when certain types of aggregate are used (in the case of ASR, this means siliceous aggregates of special crystalline type) and there must be adequate supplies of external moisture.

The total concentration of alkalis in Portland cement is expressed as the equivalent of sodium oxide (Na_2O). The cement alkalinity varies from one cement factory to another and in the UK can be in the range of 0·4–1·0%. There is also a small variation in the cement from each works. The total alkali content of the concrete generally depends mainly on the concentration in the cement and the cement content of the mix. Building Research Establishment Digest No. 258, of February 1982, suggests that a total limit of 3 kg of acid soluble equivalent sodium oxide per m^3 of concrete should prevent the development of ASR. With an alkali level of 1% in the cement, this would limit the cement content of the concrete to 300 kg/m^3. If a higher cement content was required (for reasons of strength and/or durability), then the alkali level in the cement would have to be proportionately reduced. For example, if the cement had an alkali level of 0·6%, the cement content could be raised to 500 kg/m^3. This is considering only alkalis in the cement.

It is now not unusual to find that contract specifications require that the total acid soluble equivalent sodium oxide per m^3 of concrete should not exceed 3 kg/m^3. The previous paragraph has considered the part which may be contributed by the cement and when this overall limit exists (say 3 kg/m^3) consideration must be given to alkalis in the aggregate, fine and coarse.

According to the 'Hawkins Guidelines', published by the Concrete

Society in November 1985, it is recommended that the equivalent alkali contributed by sodium chloride contamination should be ascertained. This is expressed as soluble chloride and is found by the method given in Part 4 of BS 1976 (which is now being revised and will be reissued as Part 117).

The soluble chloride thus found is then multiplied by 0·76 to give the equivalent alkali percentage by mass of the aggregate tested. For example, if the sand was found to contain 0·2% chloride ion, and the mass of sand per m³ of concrete was 500 kg, this would give 1 kg of equivalent alkali to each m³ of concrete and this figure should be added to that contributed by the cement. If the latter was 2·8 kg then the total would be 3·8 kg/m³. This would be 26% above the limit of 3% mentioned above.

The author understands the concern felt about the possibility of alkali–silica reaction occurring. However, he also hopes that the theoretical approach outlined above will not result in the wholesale rejection of aggregates which would in practice make entirely satisfactory concrete free from alkali–silica reaction.

A considerable amount of research has been carried out on ASR on a world-wide basis, and a great deal of published information is available. A few references are included in the Bibliography at the end of this chapter.

There are three serious problems associated with ASR:

(a) There is at present no really reliable short term test to determine whether a particular cement and aggregate in specific mix proportions, will or will not result in ASR in the long term.
(b) For the reaction to become visible at a general inspection, a period of up to or more than 10 years may be required.
(c) There is no known method of effecting a lasting cure.

On the other hand, the reaction may under certain favourable conditions (such as a reduced ingress of moisture) proceed so slowly that the useful life of the structure may not be substantially reduced.

From the point of view of the diagnosis of ASR it is not always easy to recognise the visible symptoms. The signs are random surface cracking (map cracking) and sometimes the exudation of a whitish gel from the cracks. Cracking is always present, but the crack pattern can be quite similar to that caused by drying shrinkage. The gel is not always present on the surface.

The presence of reinforcement will influence the crack pattern; for example, the cracking in heavily reinforced columns may well take the form of long vertical cracks. The only reliable way to diagnose ASR is to take cores and to have these subjected to microscopic examination of thin slices cut lengthwise down the core. This must be done by an experienced laboratory. Confirmatory tests include the measurement of change in length of test pieces under conditions of warmth and moisture.

Figure 8.3 in Chapter 8 shows map cracking which is typical of alkali–silica reaction.

2.3 PHYSICAL AGGRESSION TO CONCRETE

Physical aggression (wear and damage) to concrete can arise from a number of causes, the principal ones being:

1. Freezing and thawing on the outside of structures located in very exposed positions in the northern part of the British Isles and other countries with similar or more severe climate.
2. Thermal shock caused by a sudden and severe drop in the temperature of the concrete, such as spillage of liquefied gases.
3. Abrasion to concrete, such as that caused to floors in industrial buildings by steel wheeled trolleys. Similar damage can be caused to the inside of silos, bins and hoppers containing coarse granular material.
4. Damage from high velocity water. This damage can be subdivided into three types: cavitation; abrasion from water containing grit; impact from a high velocity jet.
5. Abrasion in marine structures caused by sand and shingle thrown against the structure by heavy seas and gale force winds. This is dealt with in Chapter 10.

2.3.1 Freezing and Thawing
Concrete structures which are located in very exposed positions in the north of the British Isles and other countries with a severe climate, may suffer damage by surface scaling of the concrete due to frost action. This disruption of the concrete surface is caused by the penetration of the surface layers by moisture when this is followed by sub-zero temperatures. The consequent freezing and expansion of the water absorbed in the concrete can cause disintegration of the surface layers.

Such conditions with resultant damage are fortunately rare in the UK.

It should be noted that similar trouble does occur to marine structures in extreme northern and southern latitudes.

Suitable repair measures are difficult to devise, but concrete and mortar used for repairing such damage should be air-entrained. Some basic information on air-entraining admixtures is given in Chapter 1.

Hydraulically pressed precast concrete products have proved their ability to resist the disintegrating effects of freeze-thaw, but it would only be in a limited number of special cases that they could be used for the repair of a concrete structure.

2.3.2 Thermal Shock

Damage to concrete by thermal shock is rare and occurs only in a few special cases such as the spillage onto the concrete of liquefied gases. The gases involved are those used in various industrial processes and include methane, oxygen, nitrogen and hydrogen. The temperatures of these liquefied gases are approximately $-160°C$, $-183°C$, $-196°C$ and $-253°C$ respectively. Generally the damage is confined to floors but in many cases these floors are subjected to moving loads and therefore the effect of the damage can be more serious.

The author knows of no method of repair to the concrete which can be relied upon to give a long life, and therefore the need for fairly frequent repair must be accepted. The use of high quality air-entrained concrete made with either a limestone aggregate or an artificial lightweight aggregate is likely to give the best results.

2.3.3 Abrasion of Concrete

The abrasion of concrete such as occurs on industrial floors and in silos, bins and hoppers containing certain types of granular material, is a subject on which there is a lack of precise information and knowledge. It has been established that the quality of the concrete in terms of compressive strength is the most important factor in determining the abrasion resistance of the concrete surface. The type of finish imparted to the surface is also important, but much less is known about the relative influence of different methods of surface finishing on abrasion. Most problems arise with industrial floors. The use to which the floor is put varies greatly from one industry to another.

The aggregate does have some influence on abrasion resistance but is only significant if the aggregate quality is poor or especially good. The use of certain types of limestone can result in polishing and slipperiness.

Detailed information on methods of repairs to concrete floors is given in Chapter 6.

2.3.4 Damage to Concrete by High Velocity Water

Damage by high velocity water can be divided into three main categories: cavitation; abrasion from water containing grit etc.; impact from a high velocity jet.

Cavitation

The factors involved in abrasion and impact are fairly well understood but the exact causes of cavitation have not been completely resolved in spite of the research which has taken place in various countries.

A simple, and consequently incomplete explanation of cavitation follows. With high velocity flow, i.e. exceeding about 15 m/s, depressions and irregularities in the boundary surface can cause cavitation eddies to form. This is particularly liable to occur downstream of boundary discontinuities. Figure 2.4 shows cavitating eddies downstream of a cylinder located in the cavitation test rig at Imperial College, London. It should be noted that no surface is absolutely smooth, and with very high velocities and under certain special conditions, severe cavitation damage can occur even to steel and other metals. If the absolute pressure at points of surface irregularities approaches the vapour pressure of the water, minute bubbles will form and quickly collapse. The collapse of the bubble wall can produce minute water jets having extremely high velocities. The result of the collapse of the bubbles can create an intense impact wave in the form of a series of hammer-like blows in extremely rapid succession. The effect is very destructive to high strength concrete, and as stated above, even to metals. Readers who wish for a more detailed explanation and discussion on cavitation should refer to the Bibliography at the end of this chapter.

It is obvious that if the formation of the cavitation eddies and bubbles can be prevented, then cavitation will not occur. Generally, damage by cavitation occurs on the surface of spillways, penstocks, aprons, energy dissipating basins, and syphons and tunnels carrying high velocity water.

Cavitation damage to concrete can be fairly easily distinguished from normal erosion, as the damaged areas have a jagged appearance, while erosion from grit laden water results in a comparatively smooth surface. Figure 2.5 shows typical damage to concrete by cavitation in a test rig at Imperial College. In practice, in structures such as spillways, cavitation damage can take the form of the tearing out of large pieces of concrete.

FIG. 2.4. Equipment used for research on cavitation, showing cavitation eddies forming and collapsing (courtesy: M. J. Kenn, Imperial College, London).

Abrasion from Water Containing Grit

High quality concrete is very resistant to damage by abrasion from fast flowing water containing grit. The factors which decide whether any significant damage will occur and the degree of damage, depend on a number of factors of which the following are the more important:

Fig. 2.5. Damage to concrete by cavitation in test rig (courtesy: M. J. Kenn, Imperial College, London).

1. The quality of the concrete in terms of compressive strength, cement content and wear resistance of the fine and coarse aggregate.
2. The velocity of the flowing water.
3. The quantity of the grit carried and the abrasive characteristics of the grit.
4. The flow characteristics, that is, whether the flow is continuous or intermittent and the extent to which the quantity of grit varies, hour to hour, day to day etc.

As far as the author is aware, there is no published information which shows in practical terms how the above four factors are interrelated. In the circumstances, designers have to use their engineering judgement when deciding whether special precautions are justified in the first instance and if so what these should be. This also applies when damage has occurred and repairs have to be carried out. Obviously the detailed method of repair and the techniques used will depend on the type of structure which has been damaged, the extent of the damage, and the period over which the erosion has taken place.

Impact from a High Velocity Jet
The effect of a very high velocity jet of water striking a concrete surface is

to erode the cement paste, resulting in the loosening of the fine and coarse aggregate, and this can lead to the boring of a hole through the concrete. This is the principle of cutting concrete with high velocity water jets. The pressure at the nozzle immediately before discharge can vary from 21 N/mm² to 41 N/mm². In fact high quality concrete has proved to be very resistant to erosion by clear water, but it has not been possible so far to fix any limiting factors with regard to critical velocities and concrete strengths. When erosion by a jet has occurred, probably the best solution is the provision of a steel plate securely fixed to the concrete wall or floor. The fixing detail needs careful thought and the use of epoxide mortar is generally desirable.

In cases where the damage is slight and has taken place over a fair period of time, normal concrete repair techniques can be used which are described later in this book.

2.4 CORROSION OF METALS IN CONCRETE

The metals which will be considered here are used for reinforcement or for fixings in concrete, namely, mild and high tensile steel, stainless steel, phosphor-bronze and gunmetal. Brief comments are also given on aluminium and copper as these are often fixed in or in direct contact with concrete.

2.4.1 Dissimilar Metals in Contact

When considering the corrosion of metals in concrete, a basic principle is that when dissimilar metals are in contact, corrosion of one of the metals is likely to occur. One metal will form the anode and the other the cathode; the anode corrodes and the cathode is protected. This is the principle of cathodic protection which is discussed in Chapter 4.

The problem can arise with fixings and anchors which may be in contact with steel reinforcement. It is therefore important to know which metal is vulnerable. Detailed information on this can be obtained from British Standards Institution publication PD 6484—Commentary on corrosion at bimetallic contacts and its alleviation. From this publication it will be seen that steel is anodic to copper, phosphor-bronze and gunmetal, and in some cases, to stainless steel. This means that steel reinforcement in contact with these metals is likely to corrode. The extent of the corrosion is likely to be governed by the moisture content of the concrete and the amount of oxygen at the surface of the steel.

While the principle of this type of corrosion should be known to designers, its possible effect in practical terms should not be exaggerated.

2.4.2 Mild and High Tensile Steel

It is known that steel embedded in concrete is protected against corrosion, but the reasons for this, and the limitations on this protection are often not well understood. The pore fluid in hydrated cement paste (the matrix in concrete and mortar) is strongly alkaline and has a pH in the range 12·5–13·5. This high alkalinity passivates the steel. BRE Digest 263 states that the limited oxidation of the surface of the steel which occurs within this pH range has no practical ill effect on the steel nor on the surrounding concrete.

The corrosion of steel, however small this may be, is an electrochemical process, and this activity can be measured and thus used to provide information on the likely amount of corrosion which is taking place. This fact can be used in investigations of deterioration of reinforced concrete and is discussed in Chapter 3. If the alkalinity of the concrete in contact with the steel is reduced, for example by carbonation (see Section 2.1.3 of this chapter), then in the presence of oxygen and moisture corrosion will occur.

The presence of chlorides in the concrete can stimulate corrosion even when the pH is high. The effect of chloride ions on the corrosion of steel in concrete is much more complex than is sometimes realised. A high percentage of the total chloride ion in concrete reacts with the tricalcium aluminate (C_3A) to form compounds which effectively prevent the chloride ions from attacking the steel. The balance of the chloride ions remain as 'free' chloride in the pore water of the concrete. It is the 'free' chloride ions which attack the steel but it is not possible to detect the concentration of 'free chloride' in the concrete as distinct from the total chloride.

In the UK the controlling recommendations for limiting chlorides in reinforced concrete are contained in Code of Practice BS 8110—The Structural Use of Concrete. The Code recommends that the chloride ion concentration in reinforced concrete made with ordinary or rapid hardening Portland cement should not exceed 0·4% by weight of cement. In concrete made with sulphate-resisting Portland cement, the permitted concentration of chlorides is reduced to 0·2% by weight of cement. For prestressed concrete and structural steam cured concrete, the concentration is further reduced to 0·1% by weight of cement.

In view of the above, it is the opinion of the author that detailed discussions on threshold levels for chlorides in concrete make little technical sense.

2.4.3 The Protection of Reinforcement by Concrete

From the previous Section it can be seen that the concrete surrounding the reinforcement protects the steel by passivating it unless the passivation is broken down. Even then, for significant corrosion to occur, there must be both excess moisture and oxygen.

It is normal to find that exposed concrete inside a building is more deeply carbonated than the external concrete, but that corrosion of rebars on the inside face of the member is less severe than on the outside. It must also be realised that cracks in concrete, particularly those which extend down to the reinforcement, will stimulate the early occurrence of corrosion.

Extensive research work in various countries has shown that crack width as such has little effect on the rate of corrosion in the long term, i.e. say ten years and longer. In the short term however, crack width is important. In the UK, the most important survey and investigation into the corrosion of steel in concrete was carried out in the early 1970s under the Concrete in the Oceans programme for the Department of Trade in connection with the North Sea oil structures. Some references are included in the Bibliography at the end of this chapter.

The deterioration of concrete due to the corrosion of the reinforcement may be described as following the stages set out below:

1. Passivation after the placing of the concrete.
2. Reduction and final destruction of the passivation.
3. The initiation of active corrosion and formation of rust.
4. The expansion of the rust resulting in the cracking of the concrete cover to the rebars.
5. As the rust continues to form, the cracks are extended in width and length and pieces of concrete become unstable and eventually fall off (spalling).

These stages in the deterioration of the concrete (and the reinforcement) are fundamental to the investigation of defects and are discussed again in Chapter 3.

2.4.4 Stainless Steel, Phosphor-Bronze and Gunmetal

Brief information has been given on these metals in Chapter 1. All three metals are used for fixings and anchors in reinforced concrete and are very resistant to corrosion in the type of environment likely to be met in normal building structures. In structures used for holding aggressive liquids, and in industries where aggressive vapours are generated, expert advice should be obtained from the suppliers of the fixings and anchors.

The danger of bimetallic corrosion has been briefly discussed in Section 2.4.1.

The coefficient of thermal expansion of stainless steel is about 40% greater than that of concrete.

2.4.5 Aluminium and Copper

These two metals are often embedded in concrete or are in direct contact with it. Aluminium (if not anodised or properly coated) which is in direct contact with concrete should always be protected unless it can be ensured that moisture will not be present at the contact surfaces. The protection can be in the form of bituminous paint or other paint which is alkali-resistant. The object is to form a protective barrier between the highly alkaline concrete and the aluminium.

Copper is not attacked by concrete unless the concrete contains chlorides, or ammonium compounds can gain access to the copper. The latter may occur from certain organic based adhesives.

The coefficients of thermal expansion of aluminium and copper are much greater than that of concrete.

2.5 THE USE OF BRACKISH (SALINE) WATER FOR MIXING CONCRETE AND MORTAR

In the so called developed countries of the world, it is usual to use drinking water for mixing concrete, but in the 'developing' countries it is sometimes found that drinking water is in such short supply that saline water and sea water have to be used, often with salt contaminated aggregates. The obvious question then is what effect will this have on the durability of the structure.

In the first place, HAC should not be used for concrete and mortar if the gauging water is highly saline. The manufacturers of this cement should be consulted in all cases where normal fresh (drinking quality) water is not available for mixing.

The following discussion refers to Portland cement concrete and mortar. The total dissolved solids (salts) in Atlantic water is about 32 000 ppm

FIG. 2.6. Severe corrosion of rebars in concrete column in Middle East, caused by the use of chloride contaminated aggregates (courtesy: A/S Scancem, Norway).

(mg/litre). Of these about 2500 ppm are sulphates (mostly magnesium sulphate), and 18 000 ppm are chlorides (principally sodium chloride). It has been established that the sulphates in normal Atlantic water used for mixing Portland cement concrete and mortar do not result in any significant reduction in the strength and impermeability of the concrete in the long term. The effect of the chlorides is to accelerate the setting and hardening of the cement.

The overall effect of the 32 000 ppm of salts in the mixing water would be to produce considerable efflorescence on the exposed surfaces of the concrete members. Sodium chloride is very aggressive to ferrous metals, and therefore a detailed consideration of the effect of the chloride concentration may be of interest.

As previously stated, the chloride content of Atlantic water is about 18 000 ppm; in the Persian Gulf, Red Sea, etc., the concentration is likely to be about 22 000 ppm or 2·2% by weight. If the w/c ratio of the mix is 0·45, then the chloride content of the mix expressed as a percentage of the weight of the cement would be:

$$0·45 \times 2·2 = 0·99\%, \text{ say } 1\%$$

This is more than double the recommended limit in BS 8110.

It has been noted that in those areas where it may be necessary to use sea water for mixing, the aggregates, particularly the sand, may also contain an appreciable quantity of salt. From this it can be seen that there is serious danger of corrosion of the reinforcement if sea water is used for mixing the concrete. In these circumstances it is the total sulphate and chloride content of the mix which is important.

The problem of chlorides in concrete and their effect on steel reinforcement has been discussed in Section 2.4.2. It is advisable to keep in mind the recommended maximum chloride ion concentration mentioned in that Section, and to see what practical steps can be taken to reduce the chloride content to that level.

A high sulphate content is also difficult to deal with as the sulphate will be in solution in the porewater and so will be in intimate contact with the cement particles.

While recommendations for maximum sulphate concentration vary, it is generally accepted that this should not exceed about 4·0% by weight of ordinary Portland cement; this could be increased to 6% for sulphate-resisting Portland cement when the C_3A content is limited to a maximum of 3%. The author has not seen any published research work which confirms these restrictions, but they are generally accepted as prudent.

Figure 2.6 shows severe cracking of a concrete column in the Middle East due to corrosion of rebars by chloride contaminated aggregates.

BIBLIOGRAPHY

AMERICAN CONCRETE INSTITUTE, *Corrosion of metals in concrete*, ref. SP-49, 1975, p. 142.

AMERICAN CONCRETE INSTITUTE, *Durability of concrete*, ref. SP-47, 1975, p. 390.

AMERICAN CONCRETE INSTITUTE, *A guide to the use of waterproofing, damproofing, protective and decorative barrier systems for concrete*, Report No. ACI 515.IR.79, *Concrete International*, November 1979, 41–81.

AMERICAN CONCRETE INSTITUTE, *Sulfate resistance of concrete*; six papers; ref. SP-77, 1982, p. 104.

AMERICAN CONCRETE INSTITUTE, *Erosion resistance of concrete in hydraulic structures* (reaffirmed 1979), ACI Committee 210, ref. 210R-55, p. 10.

AMERICAN CONCRETE INSTITUTE, *Guide to durable concrete* (reaffirmed 1982), ACI Committee 201, ref. 201.2R-77, p. 37.

AMERICAN SOCIETY FOR TESTING AND MATERIALS, *Standard Method of vibratory cavitation erosion test*. Designation G32–72.

BEEBY, A. W., *Concrete in the oceans; cracking and corrosion*, Report No. 2/11, August 1976 (for Department of Industry), p. 170.

BRITISH STANDARDS INSTITUTION, *Commentary on corrosion at bimetallic contacts and its alleviation*, PD 6484: 1979, p. 26.

BRITISH STANDARDS INSTITUTION, *The structural use of concrete*, BS 8110: 1985, Parts 1 and 2 (replaces CP 110).

BUILDING RESEARCH ESTABLISHMENT, *Estimation of thermal and moisture movements and stresses*, Part 2; Digest 228, Aug. 1979, p. 8.

BUILDING RESEARCH ESTABLISHMENT, *Concrete in sulphate-bearing soils and ground waters*, Digest 250, June 1981, p. 4.

BUILDING RESEARCH ESTABLISHMENT, *Alkali aggregate reactions in concrete*, Digest 258, February 1982, p. 8.

BUILDING RESEARCH ESTABLISHMENT, *The durability of steel in concrete*, Parts 1, 2 and 3. Digests 263, 264 and 265, July to Sept. 1982, p. 24.

CEMENT & CONCRETE ASSOCIATION, The Hawkins Working Party Report on alkali–aggregate reaction; minimising the risk of alkali–silica reaction— Guidance Notes. Published by the Association, no. 97.304, Sept. 1983, p. 8.

CLARK, R. R., Bonneville dam stilling basin. *J. Am. Conc. Inst.* (52), April 1956, 821–7.

CONCRETE SOCIETY, *Changes in cement properties and their effects on concrete*, Report of a Working Party, Sept. 1984, p. 22.

CONCRETE SOCIETY, *Alkali–silica reaction*. Papers given at a one-day conference, London, November 1985.

EVERETT, L. H. AND TREADAWAY, K. W. J., *Deterioration due to corrosion in reinforced concrete*, Building Research Establishment Inf. Paper IP.12/80, April 1980, p. 4.

FOOKES, P. G., COLLIS, L., FRENCH, W. J. AND·POOLE, A. B., *Concrete in the Middle East*, Five Papers. Cement & Concrete Association, ref. 12.077, p. 24.

FORBES, C. E., STEWART, C. F. AND SPELLMAN, D. L., *Snow and ice control in California*, Paper at International Symposium on snow removal and ice control—CRREL, Hanover, New Hampshire, US, April 1970.

GUTT, W. H. AND HARRISON, W. H., *Chemical resistance of concrete*, Building Research Establishment Current Paper CP.23/77, May 1977, p. 4.

HAIGH, I. P. (Ed.), *Painting Steelwork*, Construction Industry Research and Information Association Report 93, 1982, p. 124.

HOBBS, D. W., *Influence of mix proportions and cement alkali content upon expansion due to alkali–silica reaction*, Cement & Concrete Association, London, Tech. Report 534, June 1980, p. 31.

HUGHES, B. P. AND GUEST, J. E., Limestone and siliceous aggregate concrete subjected to sulphuric acid attack. *Mag. Conc. Res.*, **30** (102), March 1978, 11–18.

KENN, M. J., *Factors influencing the erosion of concrete by cavitation.* CIRIA Tech. Note (1), London, July 1968, p. 15.

KENNERLEY, R. A. AND ST. JOHN, D. A., Reactivity of aggregates with cement alkalis. *Proc. of National Conf. on Concrete Aggregates*, Hamilton, New Zealand, June 1969, pp. 34–47.

LEA, F. M., The action of ammonium salts on concrete. *Mag. Conc. Res.*, **17** (52), Sept. 1965, 115, 116.

MCCONNEL, D., MIELENZ, R. C., HOLLAND, W. Y. AND GREEN, K. T., Cement aggregate reaction in concrete. *J. Am. Conc. Inst.*, **44** (2), October 1947, 93–128.

NEVILLE, A. M., *Properties of Concrete*, 3rd ed., Pitman Books Ltd, London, 1983.

RAUEN, A., Influence of de-icing chemicals; Paper in Session 12, *International Colloquium on the Frost Resistance of Concrete*, Vienna, June 1980, pp. 31–5.

SOMERVILLE, G., *Engineering aspects of alkali–silica reaction*, Cement & Concrete Association, Interim Technical Note 8, Oct. 1985, p. 7.

TAYLOR WOODROW RESEARCH LABORATORIES, *Concrete in the oceans, marine durability survey of the Tongue Sands Tower*, Technical Report 5 (for Department of Energy), 1980, p. 141.

WILKINS, N. J. M. AND LAWRENCE, P. F., *Concrete in the oceans, fundamental mechanisms of corrosion of steel reinforcements in concrete immersed in seawater*, Technical Report 6 (for Department of Energy), 1980, p. 55.

Chapter 3

Investigation and Diagnosis of Defects in Concrete Structures

It has been emphasised in the Preface and the Introduction to this book that the investigation of defects in reinforced concrete structures should be undertaken by a qualified civil or structural engineer who has had experience in this field. The essential factor in such an investigation is the ability to recognise at an early stage the likely causes of the defects and to direct the investigation accordingly. Special attention must be given to the possibility of structural weakness.

The author's experience is that there is a real need for a regular system of inspection of all reinforced concrete structures so that any deterioration can be detected and recorded in its early stages, and a decision taken on what remedial work, if any, should be carried out.

A recommendation that inspections should be made every 3–5 years is contained in Clause 701 of Code of Practice CP 114 (1965 edition). The author feels it is unfortunate that the principle of carrying out such regular inspections was not included in BS 8110—Code of Practice for the Structural Use of Concrete. It is interesting that the FIP (International Federation for Prestressed Concrete) Working Group for the maintenance, repair and strengthening of existing structures, in their draft 'state of the art' Report in 1982, devoted a whole section to the inspection of existing structures. It should be noted that the FIP Working Group dealt with reinforced as well as prestressed concrete. The object of periodic inspections is to detect deterioration in its early stages and it in no way implies that such structures are specially vulnerable to structural failure. It is only when remedial work is neglected over a long period that the possibility of structural failure may arise.

Chapter 8 of this book deals with repairs to bridges, and it is noted in the Introduction to that Chapter that the Department of Transport require that all road bridges should be inspected at intervals not exceeding three years. The author feels that the failure of BS 8110 to make any recommendation for regular inspection of reinforced concrete structures is difficult to justify.

The course which has to be followed is:

Investigation (preliminary and detailed).
Diagnosis (based on the results of the investigation).
Preparation of Report, and when required, preparation of specification and contract documents for remedial work.

The recommendations in this chapter refer to defects in completed structures and are intended as a general guide to good practice. The form that the investigation takes will clearly depend on the type of structure and its condition.

Failures and defects in concrete structures can be placed in five main categories:

1. Structural deficiency resulting from such causes as error in design, errors in construction, impact, explosions, and change of use resulting in higher loading than was allowed for in the original design.
2. Fire damage; this can result in some weakening of parts of the structure, as well as physical damage to columns, beams, slabs, etc.
3. Deterioration due to poor quality concrete, inadequate cover to reinforcement, the presence of chlorides and/or sulphates in the concrete.
4. External chemical attack on the concrete and/or reinforcement.
5. Physical damage caused by the use to which the structure or part has been put or subjected to, such as abrasion of a floor slab in a factory, or abrasion of marine structures by sand and shingle.

3.1 THE PRINCIPLES OF THE INVESTIGATION

3.1.1 Discussions with the Client
There should be complete disclosure by both parties and the Engineer should keep the client informed on all stages of the investigation. The client should be asked to provide the Engineer with a clear brief on what he requires. This is particularly important if the client is considering

litigation with the object of recovering the cost of any remedial work which may be found necessary. If the client has this in mind he should be advised by the Engineer to seek legal advice at an early stage from a firm of solicitors experienced in construction disputes.

If litigation is not contemplated then the Engineer has to satisfy himself by the investigations that the diagnosis and the remedial work proposed are technically sound and practical.

On the other hand, if the client seeks reimbursement from others for the cost of the remedial work and possibly for consequential economic loss as well, then the Engineer will have to satisfy the Court, in the face of strenuous opposition, that all his recommendations are fair and reasonable and have taken into account his client's duty to mitigate the loss. In many instances this can result in the need for a much more detailed investigation than would otherwise be needed. It should be kept in mind that diagnosis and recommendations for remedial work are very much a matter of personal opinion which should be based on experience, sound technical knowledge, and adequate investigation.

3.1.2 Background Information
The Engineer should obtain as much information as possible about the structure:

> Date of construction.
> Details of construction (if available).
> Present use, and any previous changes of use.
> When deterioration was first noticed and whether any repairs had been carried out, and with what result.
> If the structure is post World War II construction, the names of the professional firms involved in the design and the contractor(s) who built it.

Careful consideration of the above information will often indicate possible underlying causes of the deterioration and help in drawing up proposals for the detailed investigation, sampling and testing. These are dealt with in the next Section. Mention will be made of various techniques for use in the investigation. Apart from the taking of cores, these are all non-destructive methods (NDT). Such methods have their limitations, and generally, a single technique should not be used on its own to arrive at a final decision of importance. Efforts should always be made to obtain confirmation by other means.

Some of the techniques were developed in the USA, mainly in connection with the vast amount of damage done to concrete bridges by corrosion of reinforcement due to ingress of chlorides from de-icing salts.

3.1.3 Preliminary Inspection

Ideally, the information that was suggested as needed in Section 3.1.2 should be in the hands of the Engineer before he makes his inspection, but it is very seldom that this is the case. Such information is hard to come by and in many instances is just not available. The Engineer then has to make his first inspection on the basis of the client's complaint ('pieces of concrete falling off', 'cracking', 'rust staining', 'water penetration', etc.).

Depending on the type of structure and alleged defects, the Engineer may take with him binoculars, a Schmidt Rebound Hammer, a bolster and heavy hammer, a simple means for approximate measurement of crack widths, and possibly a cover meter. The problem of access can impose severe restrictions on the amount of investigation possible at this stage.

The object is to obtain a general picture so that decisions can be taken on the following matters:

(a) Sampling and testing programme.
(b) Means of access for this programme.
(c) The need or otherwise for additional background information; for example, if structural (load-bearing) deficiency is suspected, access to the original drawings could save a considerable amount of expense in load testing or on site investigations to determine the amount and location of the reinforcement.

The taking of a few carefully selected photographs is always advisable.

The preliminary inspection may show that the damage is not structural (that is, there is no structural weakness), and the structure does not require strengthening. A discussion on an investigation involving structural deficiency is given in Section 3.5.

3.1.4 Detailed Inspection, Sampling and Testing

It is unlikely that detailed proposals for repair of a deteriorated concrete structure could be prepared on the basis of the preliminary investigation described in Section 3.1.3. It may have been possible to take some samples at this initial inspection, and tests on these for chlorides, depth of carbonation and, in some cases, sulphates, may yield useful and indicative information for use in the planning of the detailed investigation.

The first step would be the setting up of a sampling and testing programme and this is best done in consultation with an experienced commercial testing laboratory. After a joint inspection and discussion, the Engineer should prepare a clear brief for the laboratory, detailing the amount of sampling and the location of the samples, and the amount and type of testing he requires. The laboratory would normally give an estimate of the cost of the work including the cost of provision of necessary access. Alternatively, the Engineer may deal with the quotations for cradles/scaffolding himself.

The orders for the sampling and testing and for the access may be issued by the client on the recommendation of the Engineer or the whole matter may be left in the hands of the Engineer. If structural deficiency is suspected, it is advisable for this aspect to be dealt with as quickly as possible. Some general comments on this are given in Section 3.5.

The immediate problem is to decide on the amount and location and type of samples. This will vary from job to job and so only general recommendations can be given, using a 'typical' building structure as an example, with no litigation involved. For the purpose of this example it is assumed that the building was erected during the late 1960s and consisted of an insitu reinforced concrete frame (columns, beams and floor and roof slabs) with precast non-load bearing external wall panels. The initial inspection showed that the insitu and precast concrete displayed cracking, spalling and rust staining, but the precast panels were in a worse condition than the insitu concrete. This suggested that there may be a 'chloride problem' with the precast panels.

At the initial inspection, samples could only be taken at ground level due to lack of access. These samples were checked for cement content, and chlorides, and depth of carbonation was measured on site and by breaking some of the samples. Laboratory tests showed that the samples of insitu concrete had a rather low cement content compared with present day Code recommendations, but the chloride concentration was 0·4% and less by weight of cement. The samples from the precast concrete gave a different picture, a higher cement content and a high chloride content (chlorides 0·9–1·8% by weight of cement). The carbonation depth in the insitu concrete was generally greater than in the precast concrete, and in some locations had reached the rebars.

A more detailed investigation was then carried out using cradles. This consisted of an extensive cover meter survey, sampling for chlorides (single samples from 10% of the precast units, as recommended in Building Research Establishment Information Sheet IS 13/77), and random samples from the insitu concrete. A copper–copper sulphate half-cell

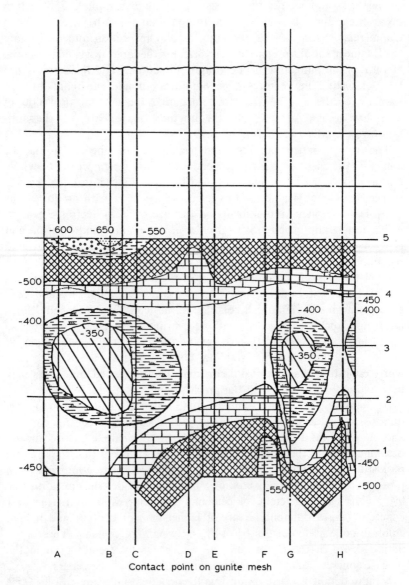

FIG. 3.1. Half-cell interpretation map-contact point on mesh in gunite (courtesy: Technotrade Ltd).

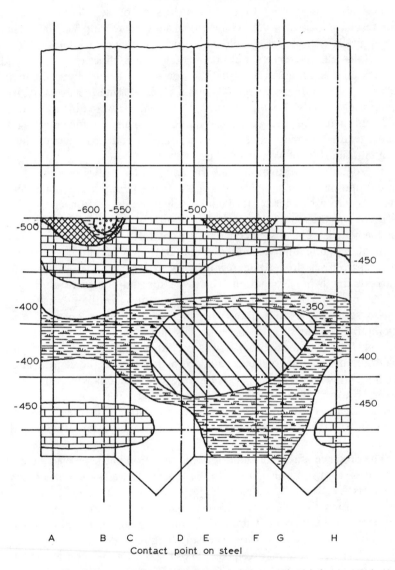

FIG. 3.2. Half-cell interpretation map-contact point on main reinforcement in the concrete below the gunite (courtesy: Technotrade Ltd).

survey was carried out in a few selected areas where there were no signs of deterioration. In one of these areas, the soffit of an external staircase, the results indicated a 95% chance that active corrosion was occurring (readings in excess of 350 millivolts). For more detailed information on the 'Half-Cell', reference should be made to the Bibliography at the end of this chapter. Figures 3.1 and 3.2 show half-cell interpretation maps relating to corroding steel on a diagonal brace of a marine structure; it is interesting that the Testing Laboratory responsible were able to produce different maps for the gunite layer and the main reinforced concrete underneath. The maps and information are reproduced by courtesy of Technotrade Ltd of Welwyn Garden City.

The detailed investigation and sampling from cradles confirmed the initial findings from samples at ground level; the important point being that the precast concrete did contain a high concentration of chlorides and this was the main cause of the corrosion of the nominal reinforcement in the panels. The insitu concrete was of rather low quality (cement content variable and generally about $280 \, kg/m^3$), chlorides probably present from sea dredged aggregates, very variable cover (5–50 mm), and variable depth of carbonation. There was no sign of structural distress.

The presence of the high chloride concentration in the precast panels raised serious questions about the long term durability of any repair method. The cement used in the panels had a rather low C_3A content (about 7%). This meant that it was probable that the amount of 'free' chloride ions in the concrete would be higher than if the cement had a high C_3A content (say 11%). It is the 'free' chloride ions which attack the steel. However, it should be kept in mind that when concrete becomes saturated, the chloride ions which have combined with the C_3A in the cement will tend to become 'free' and be available for attacking the rebars.

The presence of the high chloride concentration in the precast panels is of course a very serious matter. In the US and Canada work has been in progress for about 20 years to find an effective way to deal with high concentrations of chlorides in reinforced concrete, mainly salt contaminated bridge decks. The result to date appears to be that the most practical and effective method to ensure long term durability is cathodic protection of the reinforcement. The break-through came with the development of continuous anode systems. There is a brief description of the principles of cathodic protection and its application to reinforced concrete repairs in Chapter 4, Section 4.6, and additional information is given in Chapter 8. At the time this book was written, a number of reinforced concrete buildings were being repaired in the UK using cathodic protection to ensure long term durability of the reinforcement in chloride contaminated concrete.

The client was informed of the situation by means of a detailed Report from the Engineer which included the test reports from the commercial testing laboratory.

At that time there was no site experience with the use of cathodic protection for existing concrete structures, and so it had to be accepted that either the precast panels had to be removed and replaced or repaired in the knowledge that further repairs may be necessary in 5–10 years time. The latter course was selected by the client.

3.2 THE CAUSES OF CRACKS IN REINFORCED CONCRETE

While the deterioration of reinforced concrete structures is usually accompanied by cracking and spalling, it is important to realise that cracks as such are not necessarily defects needing repair.

It is important to determine the cause of the cracks, while their width, position and direction, and the degree of exposure, are all matters which will help to decide what action if any should be taken.

The cause of the cracking can only be decided by a careful investigation, but their width, position in the member and the degree of exposure can all be easily determined. A question which follows from the above is whether there is an 'acceptable' crack width. By this is meant a width which, under the given circumstances of exposure, can be considered as insignificant and therefore needs no repair or treatment. There is no precise and generally agreed answer to this.

British Standard Code of Practice, BS 8110: Part 2, The Structural Use of Concrete, recommends a maximum crack width of 0·3 mm for general conditions. Crack widths are measured at the surface of the concrete and it is assumed that they reduce in width fairly rapidly as the crack extends inwards. This assumption is no doubt correct when the cracks are caused by tension in the tension zone of a stressed member, but may not hold for cracks caused by thermal contraction and plastic settlement. It is very difficult to measure the width of a crack with any real degree of accuracy under site conditions.

An important practical question in connection with acceptable crack widths is how long that width will be maintained. Will the crack tend to close up or to gradually widen with the passage of time? When the crack is caused by some temporary overload it will close again when this excess stress is reduced. However, when the cause is permanent then it is likely that the crack will not close and under external conditions there will be a definite tendency for the width to increase with age.

Beyond a certain width, moisture is likely to penetrate into the crack, even for a short distance, and then if the temperature falls below $0°C$, this moisture will turn into ice and the resulting expansion will tend to cause spalling along the edges of the crack. This will occur every time the freeze-thaw cycle is repeated. The crack will then gradually widen and eventually moisture will reach the reinforcement and corrosion will start. The corrosion products (rust) occupy a larger volume than the original metal and the expansive forces will crack and spall the concrete.

Certain factors may operate to counteract the widening of cracks described above. One is compressive stresses which may develop at right angles to the line of the crack, such as occur in the compression zone of structural members. Another is the leaching of lime from the cement paste caused by the movement of moisture through the concrete and the continued hydration of cement particles. This phenomenon is often referred to as 'autogenous healing'. A third is the slight expansion of concrete which occurs from what is sometimes called 'reverse drying shrinkage' or moisture movement of the concrete itself. A typical example is the walls and floor of a water retaining structure after it is filled and put into operation.

A considerable amount of research work has been carried out in various countries on the effects of crack width on the corrosion of steel in concrete. This work was admirably summarised by A. W. Beeby in a paper published in 1978 (see Bibliography at the end of this chapter). The conclusions were that in the short term, say up to two years from construction, the width of the cracks (which were induced in the test specimens by loading) will significantly influence the amount of corrosion on the bar where it intersects the crack. However, in the long term, say 10 years, the influence of the crack width on the amount of corrosion is negligible. The reasons for this are examined in some detail in the paper. However, in considering corrosion of steel in concrete, it must be remembered that this aspect of corrosion relates only to corrosion initiated by cracks which intersect the rebars. When corrosion is initiated by other factors, such as chlorides in the concrete, carbonation and porous concrete, cracking will occur due to the expansive forces created by the formation of the corrosion products (rust). In the cases discussed by Beeby there was cracking resulting in corrosion; in most cases of deterioration it is the corrosion which causes the cracking.

3.2.1 Types of Cracks

There are many causes of cracking in reinforced concrete structures, but for practical purposes, these can be placed in two main categories:

1. *Structural cracking*. This would indicate that the structure, or far more likely, one part of the structure, was showing signs that the time was approaching, or had been reached, when it could not safely support the loads to which it was being subjected.

   This state of affairs may be brought about by:

   (a) error in design;
   (b) loading in excess of the design load even allowing for the built-in factor of safety;
   (c) some error or shortcoming in construction (workmanship and/or materials);
   (d) physical damage, explosion, impact, fire;
   (e) severe deterioration resulting in serious corrosion of reinforcement (many factors can cause this).

2. *Non-structural cracking*. This is cracking by any cause except those listed above. However, it must be kept in mind that in reinforced concrete if cracking of this type is ignored, corrosion of rebars can proceed, and after a period of time, the bars become so corroded that the member or structure becomes structurally unsafe. In other words, this is cracking in a structure or part of a structure, which at the time of inspection is structurally sound and possesses an acceptable factor of safety, but if neglected, may result in structural distress.

A careful examination of the cracks will give valuable information and will generally indicate their cause. Such factors as their position, pattern, direction (vertical, horizontal or inclined), their direction in relation to the main reinforcement, i.e. whether parallel or transverse, are all important. Any deflexion of the member is also significant. The width of the cracks should be measured as accurately as possible, but this will vary appreciably as the crack winds its way around pieces of aggregate; a high degree of precision is neither necessary nor practical. The depth of cover to the reinforcement must also be measured.

If a reasonable assessment of the cause of the cracking cannot be made on the basis of the information obtained above, then the reinforcement drawings, calculations and the specification should be examined.

The author's experience is that the majority of cracking is non-structural and in existing buildings is caused by poor quality concrete and/or inadequate cover to the reinforcement. Plastic cracking appears very soon after the concrete has been cast. Thermal contraction cracking is usually present when the formwork is removed, or if there is no formwork, within the first 48 h of casting, although it is often not noticed for several days or even weeks. The application of a curing membrane helps to mask these fine cracks. The width of thermal contraction cracks is usually increased by normal drying shrinkage, most of which occurs within the first 3 months after casting.

3.2.2 Non-structural Cracks

Plastic Cracking
There are two categories of plastic cracking. The first, and most common, results from a too rapid evaporation of moisture from the surface of the concrete while the concrete is still plastic, and is usually referred to as plastic shrinkage cracking.

Investigations by various authorities have shown that the principal cause of plastic shrinkage cracking on horizontal surfaces is a rapid evaporation of moisture (drying out) from the surface of the concrete. When the rate of evaporation exceeds the rate at which water rises to the surface (known as 'bleeding'), plastic cracking is very likely to result..

The rate at which the water in the mixed concrete reaches the surface and the total quantity involved depends on many factors, not all of which are yet completely understood, but the following are known to play an important part in this phenomenon:

Grading, moisture content, absorption, and type of aggregate used.
Total quantity of water in the mix.
Cement content.
Thickness of the concrete slab.
Characteristics of any admixtures used.
Degree of compaction obtained and therefore the density of the compacted concrete.
Whether the formwork (or sub-base) on which the concrete was placed was dry or wet.

The rate at which the water in the mixed concrete reaches the surface will also depend on a number of factors which are much better understood, and these are:

Relative humidity.
Temperature of the concrete.

Temperature of the air.
Wind velocity.
Degree of exposure of the surface of the slab to the sun and wind.

Plastic shrinkage cracking shows itself as fine cracks which are usually fairly straight and can vary in length from about 50 to 750 mm. They are often transverse in direction. In some cases several of the cracks are parallel to each other and the spacing can vary from about 50 mm to 90 mm. The cracks are usually shallow and seldom penetrate below the top layer of reinforcement, although in severe cases they can extend to a greater depth and even right through the slab. They are generally numerous. Figure 3.3 shows typical plastic cracking in a floor slab.

It is quite common for this type of cracking to occur in hot sunny weather or on days when there is a strong drying wind, and can cause consternation for those who do not realise what it is. Unless it is very severe, when it may result in deep cracks and in a permanently weakened surface to the slab, it does no real harm.

Drying Shrinkage Cracking
The author's experience is that drying shrinkage cracking is generally confined to (a) non-structural members which have either no reinforce-

FIG. 3.3. Severe plastic cracking in a floor slab (courtesy: A/S Scancem, Norway).

ment or only 'nominal' reinforcement for handling purposes; (b) thin toppings, screeds and rendering.

In most cases it is caused by a badly designed mix, the effects of which are aggravated by inadequate curing. The use of calcium chloride as an admixture, or the presence of chlorides in the aggregates, will increase drying shrinkage.

The inadequacies in the mix design may be just too much water in the mix, or the use of poorly graded fine aggregate which contains a high proportion of very fine material. The higher the percentage of fine material in a concrete or mortar, the higher will be the water demand to obtain a given workability.

All concrete and mortar shrinks on drying out, and as stated in the previous section, drying shrinkage will tend to widen cracks which have been caused by other factors such as thermal contraction.

It is important to remember that contraction due to drying shrinkage after 90 days is about 60% of that which will take place in 360 days.

The total shrinkage (moisture movement) is made up of irreversible shrinkage and reversible shrinkage. On first drying out an appreciable amount of the total shrinkage is irreversible but after several cycles of wetting and drying, the shrinkage becomes almost entirely reversible. Tabulated figures for the two types of moisture movement for a number of the major building materials are given in Building Research Establishment Digest 228, 1979. The figures for concrete and mortar refer to unreinforced test specimens, which are, for practical purposes, unrestrained. From the figures given in the Digest it can be seen that the two components are almost equal. Shrinkage can be partly restrained and crack width and spacing controlled by the use of reinforcement.

In addition to the factors mentioned above which affect the amount of drying shrinkage, there can be that relating to the use of shrinkable aggregates. These shrinkable aggregates occur in various countries including the US and South Africa and, in the UK, in the North of England and Scotland. If these aggregates are used without the necessary precautions being taken, serious cracking and deflexion in the members can occur. Basic information on the problem and how to deal with it is contained in Building Research Establishment Digest No. 35 (second series) of 1968, which is based on many years of research.

Plastic Settlement Cracking
The second form of plastic cracking is due to settlement of the concrete in the formwork after compaction has been completed, and is thus quite

different in its origin to the plastic shrinkage cracking just described.

Plastic settlement cracking can be caused in two basic ways. The first is by the resistance of the surface of the formwork to the downward movement of the plastic concrete under the influence of the poker vibrators and the force of gravity. This resistance delays the downward movement of the concrete and so, as the concrete stiffens, a crack is very likely to form. This is a surface defect close to the formwork; the cracks do not penetrate deeper than about 20–25 mm and are wider at the surface.

With the second type of plastic settlement cracking, the cracks are caused by the concrete becoming 'hung-up' on either the reinforcement or spacers (or both) and a crack forms as the concrete stiffens. The cracks penetrate at least to the reinforcement and are usually wider inside the concrete than on the surface, and when the concrete is cut away, voids are found adjacent to the reinforcement on which the concrete has become hung-up. The author considers that in most cases crack injection is the best repair method for this type of plastic cracking and some information on the technique is given in Chapter 5.

With both of these types of plastic settlement cracking, changes in the mix proportions and revibration of the concrete a short while after the first compaction has ceased will usually cure the trouble.

Thermal Contraction Cracking
During the setting and early hardening process of concrete considerable heat is evolved by the chemical reaction between the water and the cement which results in a rise in temperature of the concrete. The actual rise, the peak temperature, and the time taken to reach the peak and then to cool down, will depend on a large number of factors of which the following are the most important:

The ambient air temperature.
The temperature of the concrete at the time of placing.
The type of formwork used (whether timber, plastic or steel), and the time the formwork is kept in position.
The ratio of the exposed surface area of the concrete, i.e. the area not protected by formwork, to the volume of concrete.
The thickness of the section cast.
The type of cement used and the cement content of the mix.
Whether any provision is made for the thermal insulation of the concrete after the formwork is removed.
The method of curing.

The above are not placed in order of importance because this order depends on site conditions.

As the temperature of the concrete rises the concrete expands and when it cools down it contracts. The coefficient of thermal expansion (and contraction), depends on a number of factors of which the principal are the type of aggregate and the mix proportions.

Unless the section (floor, wall or roof) is conpletely unrestrained (a state of affairs never met in practice), thermal stress will be developed as the concrete cools down and contracts. The greater the restraint the higher the thermal contraction stress. These stresses are generally tensile, but in certain parts of a structural member they may be compressive. These tensile stresses often exceed the strength of the concrete in tension and/or the bond strength between the concrete and the reinforcement, and then cracking will occur.

Thermal contraction cracks extend right through the member. Figure 3.4 shows typical cracks in a cantilever balcony slab. While these cracks are seldom significant structurally, they do form permanent planes of weakness through the member unless they are properly repaired. Normal drying shrinkage will tend to widen these cracks, which when first formed are usually very fine, often no wider than 0·1 mm. For this reason they are often not noticed for several weeks after casting. This type of cracking is often wrongly attributed to drying shrinkage. The latter takes place slowly under normal conditions in the UK and in the first 90 days after casting is unlikely to exceed about 60% of the maximum which will eventually occur, over a long period of time.

Cracking Caused by Bad Workmanship

There are many ways in which careless workmanship can result in the cracking of recently placed and immature concrete, but only two will be mentioned here. One is the careless removal of formwork from beams and columns and soffits of slabs. In beams and columns the arrises are particularly vulnerable to damage in this way. Another is in the formation of movement joints (expansion joints), where again the arrises are liable to damage. Figure 3.5 shows a 'scarf' type of movement joint in an external beam which was so badly made that the top nib cracked right through. Such joints are best avoided and another more simple detail used.

Cracking Caused by Alkali–Aggregate Reaction

For the purpose of this short discussion, alkali–aggregate reaction is

FIG. 3.4. Thermal contraction cracking in cantilevered balcony slab (courtesy: Professor B. P. Hughes, Birmingham University).

confined to alkali–silica reaction (ASR) (see also Chapters 2 and 8).

It is usual to find that the initial investigations were started by a report of surface cracking. Generally this cracking is found to occur many years after the structure was completed, but this is not invariable. The crack pattern varies, but in many cases it is seen to form a random distribution, known as map cracking, and is often confused with drying shrinkage, but the latter occurs in the very early history of the concrete. Figure 8.3 in Chapter 8 shows typical map cracking which, in the case photographed, was later diagnosed as alkali–silica reaction.

In structural members such as columns and beams where reinforce-

Fɪɢ. 3.5. Badly executed movement joint in concrete edge beam which had been completely covered with rendering (courtesy: Hinton Consultancy Ltd).

ment is in the form of fairly heavy main bars, the cracks often follow the lines of the bars.

In addition to the cracks, there is sometimes a yellowish gel extruded from the cracks. However, carbonation may turn this to a whitish colour so that it looks like the leaching of lime. A frequently observed condition in cases of ASR is that the concrete is wet and occurrence of ASR is usually associated with the presence of excess moisture. The really important point to remember is that definite diagnosis can only be made by a skilled laboratory examination.

3.3 DIAGNOSIS OF DEFECTS

The previous Sections discussed investigations in general, and in particular of a 'typical' reinforced concrete building structure. The question then arises as to how these results can be interpreted and a reasonable diagnosis of the cause of the deterioration arrived at. This interpretation and diagnosis will form the foundation for recommendations for repair.

3.3.1 Cracks
The cracking of reinforced concrete is such a contentious subject that a separate Section in this chapter has already been devoted to it (see 3.2). Sufficient to say here that the author's experience is that the vast majority of cracks in reinforced concrete are not structural, that is they have not been caused by lack of strength in the member to support the loads imposed on it.

In the example used in Section 3.1.4 the vertical cracks in the beams may have been initiated by drying shrinkage and the locations determined by the presence of the stirrups, particularly as many of them were close to the surface.

3.3.2 Chlorides
The concentration of chlorides in the insitu concrete suggested the use of marine aggregates, and subsequent investigations confirmed this. The concentrations were well within the recommended limit in Codes at the time the building was erected, and just about met the recommendations in the present Code, BS 8110. The level of chlorides may have aggravated the corrosion process, but were not the real cause in the insitu concrete.

In the precast concrete, the level of chlorides was higher than recommended at the time the building was constructed and was about 3–5 times the concentration which is now recommended in BS 8110. This, combined with the many instances of low cover to the rebars, made corrosion of the steel inevitable. The investigation showed that in spite of the widespread spalling of the precast units, the actual depth of corrosion of the rebars (i.e. the loss of steel and consequent reduction in diameter) was minimal; also, the units were non-load bearing.

3.3.3 Cement Content
In the insitu concrete, this was not unusual for concrete made in the middle to late 1960s, but was lower than would meet the recommendations in present day Codes. The higher cement content is needed to help ensure long term durability (see Chapter 2), and it is the comparatively low quality of the insitu concrete in terms of cement content and relatively high permeability which is the basic cause of the corrosion of the rebars.

3.3.4 Carbonation
The high permeability in the insitu concrete is confirmed by the depth of carbonation (10–25 mm), and can be compared with that of the precast concrete (3–8 mm). The effect of carbonation on the possible corrosion of rebars in concrete is discussed in Chapter 2.

3.3.5 Cover to Reinforcement
The low cover had contributed to the corrosion in both the insitu and the precast concrete members. Variations in cover can result in variations in moisture content at the surface of the steel and this itself can increase the risk of corrosion.

3.3.6 The Half-cell
This showed that corrosion was beginning (or more correctly, at the time of the survey was in a more active phase) in three out of the four locations. It was interesting to note that there were no visible signs of corrosion in any of the four locations (no cracking and no rust staining).

The diagnosis of the deterioration in the 'typical' reinforced concrete building structure may therefore be summarised as follows:

3.3.7 Insitu concrete
The deterioration was the result of the concrete having inadequate capacity to protect the reinforcement in the long term under conditions of severe exposure (i.e. exposed to driving rain, alternate wetting and drying and to freezing whilst wet—BS 8110, Table 6.2). This situation was aggravated by many instances of low cover, variations in cover, and the presence of chlorides in the mix. All these factors in combination

caused the corrosion of the reinforcement, the expansive formation of rust causing cracking and spalling of the concrete. The corrosion of the rebars had not proceeded sufficiently far to cause any reduction in the original design assumptions and there was no sign of structural inadequacy.

Repairs were needed to be carried out as soon as possible, as the corrosion would continue, and the rate of corrosion was likely to accelerate. It was anticipated that a satisfactory long term repair could be carried out to the insitu reinforced concrete.

3.3.8 Precast Concrete (non-load bearing)

The deterioration was due largely to the presence of a relatively high concentration of chlorides in the concrete, aggravated by instances of low cover and variations in the depth of cover. The concrete was basically good quality, otherwise the deterioration would have been much more serious. These facts meant that an entirely satisfactory long term repair to the precast members could not be implemented at that time.

The client must be informed of such a situation and the Engineer should do his best to prepare proposals for remedial work to both insitu and precast concrete which would give as good long term performance as possible.

3.4 THE PREPARATION OF REPORTS AND CONTRACT DOCUMENTS

3.4.1 Reports

The format of the Report and its length will obviously depend on the importance of the repair work and its magnitude. As mentioned in Section 3.1.1, the question of possible litigation should have been discussed with the client before the detailed investigation was started; this will certainly influence the formalities of the Report and preparation of the Contract documents and adjudication of the tenders.

Apart from the written section of the Report, there should be a comprehensive set of good photographs in colour, cross referenced to the Report. Elevational drawings or diagrams, showing the location of the main defects, such as cracking and rust staining, hollow-sounding areas and spalling, will help to present a clear picture of the magnitude of the deterioration.

3.4.2 Contract Documents

These are likely to consist of:

General Conditions of Contract.
Instructions to Tenderers.
Description of Works.
Specification.
Schedule of Work.

The General Conditions of Contract can be the Agreement for Minor Building Works issued by the Joint Contracts Tribunal (JCT) for work estimated to cost up to about £50 000. For larger contracts, the author suggests the Institution of Civil Engineers Conditions of Contract, 5th edition, in preference to the JCT Standard Forms of Building Contract as the latter are frequently revised and each revision is more complex than the edition it replaces.

The Instructions to Tenderers, Description of Works and the Specification should be as clear and unambiguous as possible. Regarding the Specification, the following is an extract from BSI Publication, PD 6472: 1974—Specifying the Quality of Building Mortars, and in the opinion of the author is an excellent description of how any construction Specification should be drafted:

'The specification should be drafted with a clear and realistic concept of the way it can be enforced.

This implies that it should be possible to check whether every requirement is being met and if any breach of the requirements is discovered, suitable remedial action can be demanded. The action to be taken should be related to the probable consequences of the lack of compliance with the requirements ...'

The author does not generally recommend the use of a Bill of Quantities for the type of repair work described in this book. All that is needed for the majority of contracts can be adequately covered by a Schedule of Work.

It is important to know the prices submitted by the tenderers for the various items of work; this information can be very useful when studying the tenders with a view to adjudication. The tenderers should be required to commit themselves on their intentions with regard to the use of subcontractors, and on the type and cost of the access equipment they intend to use, including protection of the public (where appropriate). It is often unwise to accept the lowest tender, even though tenders have been invited on a selected list. It is the author's strongly held opinion that for a satisfactory job, the contractor must be in a position to make a reasonable profit. If he finds he is making a loss, trouble is inevitable.

Forfeiture or Determination of the Contract will certainly result in a very large increase in the final cost to the client, and the same applies if the Contractor goes into liquidation. Unfortunately, some clients appear to hold the view that if the Contract documents are properly drawn up the Contractor can be forced to carry out the whole of the work exactly in accordance with the Contract irrespective of whether the Contractor has seriously under-priced the job. This is only correct in theory; it is better to avoid such a situation arising by a sensible adjudication of the Contract at tender stage.

The control of the Contract is best done by the Engineer. However, it is advisable for the Engineer to make it clear to the Employer what duties he is actually undertaking, as it is most unlikely that he can spare the time to 'supervise' the work. In practice, the most he can do will be to 'inspect' the work say once a week. This means that the Contractor will be completely without supervision for over 90% of the time unless the Employer engages a full-time Resident Engineer or Clerk of Works, which is highly unlikely except on major jobs, and the Employer should understand this.

3.5 INVESTIGATIONS FOR STRUCTURAL REPAIRS

In Sections 3.1.3 and 3.1.4 a 'typical' building structure was taken as an example for the investigation for non-structural deterioration. In this Section it is intended to consider briefly the situation if the preliminary inspection indicated that there was likelihood that structural repairs (strengthening) of the structure would be required.

The suspected structural weakness could result from a number of reasons of which the principal are:

(a) Error in original design or in construction (very unusual).
(b) Physical damage caused by impact, explosion or fire.
(c) Change of use resulting in an increase in loading.
(d) Serious neglect of execution of repairs necessitated by general deterioration (on the basis that if the repairs had been carried out in time structural damage would have been prevented).

Generally, the signs which indicate structural deficiency together with basic information on the structure, including change of use, will suggest the likely cause to the Engineer carrying out the investigation. Impact, explosion and fire are obvious examples.

What are the signs of possible structural distress? No precise rules can be laid down and considerable experience is needed, and the following brief notes are based on the author's experience.

Diagonal cracks in beams and walls usually denote high shear stresses and should always receive special attention. Deflexion of beams and other horizontal members suggest overstressing and should be investigated. The same applies to the bowing of vertical members such as columns and load bearing walls and panels. Deflexion is usually accompanied by cracking at right angles to the main bars, but with bowing or overload of vertical members, the cracks may be parallel to the main bars. Cracking and spalling and some degree of distortion can be caused by inadequate provision having been made for thermal movement. While these cracks may not necessarily denote structural weakness, they are considered as a type of structural defect by many engineers. To effect a satisfactory remedy in an existing building can be very difficult and in some cases virtually impossible.

In the case of suspected error in design or construction, every effort should be made to obtain the original drawings and specification for the concrete. If this information is not available, then consideration has to be given to either a loading test (or tests) as appropriate, or other means found to check the actual construction against fresh calculations. The checking of what is actually in the structure (reinforcement and concrete strength) requires very detailed consideration and it would be prudent to consider more than one method. Non-destructive testing techniques are very useful and have increased in scope and accuracy in recent years, but they must be used with discretion and with a knowledge of their limitations. The Report on the radiography of Swaythling Bridge on the A27 trunk road near Southampton showed that the location and diameter of reinforcement in concrete 870 mm thick could be accurately determined.

There is no clear dividing line between an investigation into a deteriorated concrete structure where there is reason to suspect structural distress and one where there must be a structural appraisal. It is of course the 'grey' areas which are the most difficult to deal with, particularly when an attempt is made to describe the procedure to be followed. For example, in the opinion of the author, the replacement of corroded fixings for non-load bearing concrete wall panels is a structural repair, and the same would apply to the replacement of deteriorated bearings in a bridge.

It has been mentioned in Section 3.2 that the majority of cracks in concrete structures are of non-structural origin. On the other hand it is most unlikely that there would be structural distress in a reinforced concrete structure without visible cracking. It is therefore important to realise the true significance of the cracks as early in the investigation as possible.

Again, transverse cracks at regular spacing in the surface of a suspended warehouse floor slab could be due to 'shrinkage' or they could indicate differential movement between adjacent parts of a structural floor which consisted of precast double tee beams with an insitu slab cast on top. If the spacing and location of the transverse cracks coincided with the edges of the flanges of the double tee units, this would suggest that the insitu topping may have cracked along these lines. In such a case, cores would almost certainly have to be taken on the line of some of the cracks to ascertain the depth of the cracks.

When the quality of the concrete has to be investigated, it is usual to take cores and to test these for strength and use the concrete for analysis for cement content, and if felt desirable, for chlorides and sulphates. It is most important that the location and number of samples should be such that the test results can be considered as reasonably representative of the concrete as a whole.

In case the results are questioned at some time in the future it is important that the sampling should be representative of the concrete under investigation, and can be shown to be such. Also that the testing and analysis was carried out strictly in accordance with the relevant Parts of BS 1881—Testing Concrete, or the National Standard relevant to the country in which the structure is located.

For existing structures which, for one reason or another, have deteriorated, the usual tests are for:

Compressive strength.
Mix proportions (cement content).
Concentrations of chlorides and/or sulphates.

Other tests which are sometimes required are:

Absorption.
Type of cement.
Grading of aggregates (this can only give a general idea of the grading and too much significance should not be attached to the results).

Tests for chlorides are best carried out as recommended by the Building Research Establishment Information Sheet IS 13/77 and previously discussed in Section 3.1 of this chapter.

3.5.1 Compressive Strength

The test which usually invokes the most controversy is that for the determination of compressive strength of the concrete in the structure by means of cores. A great deal has been written about this. In the UK the most authoritative publications are the Concrete Society Report No. 11, British Standards BS 6089 and BS 1881, all of which are included in the Bibliography at the end of this Chapter. The cores should be taken, prepared and tested in accordance with the appropriate National Standard (in the UK, BS 1881, Part 120:1983). Thereafter, the procedure to be followed will depend on what information is required. The test on a core will give the compressive strength of that core, from which the Actual Strength and the Potential Strength can be calculated. It is essential to realise what these really mean and to decide which is required in each particular case.

The calculated estimate of the Actual Strength will be equivalent to the cube strength of the concrete represented by the core; for all practical purposes, the actual strength of the concrete at the location where the core was cut, will be as though the sample taken had been a cube cut exactly from the concrete and of the same dimensions as a standard cube cast from the fresh concrete. It can thus be seen that the estimate of the Actual Strength is needed when it is required to assess the strength of a concrete member (beam, column, slab, etc.).

The calculated estimate of the Potential Strength will be equivalent to the standard 28 days cube strength. This is required when it is necessary to determine whether the concrete at the location from which the core was cut, had the specified 28 day cube strength; such information is needed when the cube test results are below those required by the Contract. It is normally required during or soon after the construction of a reinforced concrete structure.

The Concrete Society Report No. 11 gives detailed recommendations for the calculation of both Actual Strength and Potential Strength, while BS 6089—Guide to the Assessment of Concrete Strength in Structures, covers only the estimation of Actual Strength.

The vast majority of investigations into concrete strength in existing buildings are concerned with assessment of Actual Strength, and readers are referred to the two publications mentioned for complete recommendations for such assessment.

3.5.2 Cement Content, Original Water Content, Water Absorption
When questions of durability are involved and/or comparison of the existing concrete with the original specification is required, tests are carried out to determine, as far as this is possible, the actual mix proportions and cement content, and the original water content.

Cement Content
In considering the cement contents obtained by chemical analysis it must be remembered that the tolerance on accuracy is about $\pm 10\%$ when the work is carried out by a first class laboratory. Consideration must also be given to how representative the samples are of the concrete as a whole. Recommended cement contents under CP 114 were not directly related to conditions of exposure; this was changed with the publication of CP 110 in 1972, which Code was completely revised and published in 1985 as BS 8110, Parts 1 and 2.

Original Water Content
There is a growing tendency in the UK for carrying out tests to try to determine the original water content of the concrete when durability is in question. From discussions which the author has had with experienced cement chemists, he is doubtful about the usefulness of such a test. The method of carrying out the test is given in Section 5 of BS 1881: Part 6: 1971. The opening paragraph of this Section should be read with care. It appears that reliable laboratory work has only been carried out on concrete up to 28 days old, made under strictly controlled conditions; the accuracy then would be about ± 0.25 of the water/cement ratio. It is just not possible to say what the accuracy would be for concrete made under site conditions and which is several years old. This test is best avoided.

Water Absorption of Concrete
While it is useful to know what this is, it can be very difficult to know what the results should be compared with unless the original specification contained specific requirements including method of test. Absorption should not be confused with permeability, and reference can be made to Chapter 2, Section 2.1.2. In the UK there are no recommendations for maximum permissible absorption for structural concrete. However, there are recommendations for absorption in a number of British Standards for concrete products; these include:

BS 368: Precast concrete flags.
BS 340: Precast concrete kerbs, channels, edgings and quadrants.

BS 5911 Part 1: Concrete cylindrical pipes and fittings.
BS 4131: Terrazzo tiles.
BS 1217: Cast stone.

The general method of test for absorption is given in BS 1881: Part 122, but the Standards mentioned above generally contain their own requirements for carrying out this test.

There is also a method of test to determine surface absorption, known as the ISAT (Initial Surface Absorption Test); this is described in Section 6 of BS 1881: Part 5: 1970. This test was originally intended for precast concrete units, e.g. cast stone, and there are many difficulties in trying to use it on site cast insitu concrete.

The author will conclude this Section with a quotation from page 7 of the excellent Report issued by the Institution of Structural Engineers in July 1980—Appraisal of Existing Structures:

'... when assessing existing structures engineering judgement should take precedence over compliance with detailed clauseš of Codes ...'

The comments in the Report under Section 2.3, pages 8 and 9, on 'Responsibilities' are also very important to any engineer undertaking the structural appraisal of an existing building.

BIBLIOGRAPHY

AMERICAN CONCRETE INSTITUTE, *Temperature and concrete*, Committee Report on effect of temperature on concrete; SP.25, 1971.

AMERICAN CONCRETE INSTITUTE, *Causes, mechanism and control of cracking in concrete* (Bibliography No. 9), ACI Committee 214, ref. B-9, 1971, p. 92.

AMERICAN CONCRETE INSTITUTE, *Concrete core tests* (Bibliography No. 13), ACI Committee 214, ref. B-13, 1979, p. 30.

AMERICAN CONCRETE INSTITUTE, *Guide for making a condition survey of concrete in service* (reaffirmed 1979); ACI Committee 201, 1968, p. 16.

AMERICAN CONCRETE INSTITUTE, *Control of cracking in concrete structures*, ACI Committee 224, ref. 224R-80, 1980, p. 42.

AMERICAN CONCRETE INSTITUTE, *Strength evaluation of existing concrete buildings* (revised 1982), ACI Committee 437, ref. 437R-67, 1982, p. 6.

AMERICAN SOCIETY FOR TESTING AND MATERIALS, *Standard test method for half cell potentials of reinforcing steel in concrete*, ANSI/ASTM 876–80.

BEEBY, A. W., Cracking: what are crack widths for?, *Concrete*, July 1978, p. 31.

BICKLEY, J. A., The variability of pull-out tests and in-place concrete strength. *Concrete International* (ACI), April 1982.

BRITISH STANDARDS INSTITUTION, *The structural use of concrete*, BS8110:1985, Parts 1 and 2 (replaces CP 110).

BRITISH STANDARDS INSTITUTION, *Recommendations for non-destructive methods of test for concrete*, BS 4408, Parts 1–5.

BRITISH STANDARDS INSTITUTION, *Methods of testing concrete*, Parts 1–6 and 101–122, BS 1881.

BRITISH STANDARDS INSTITUTION, *Guide to the assessment of concrete strength in existing structures*, BS 6089.

BRITISH STANDARDS INSTITUTION, *Methods of testing mortars, screeds and plasters*, BS 4551.

BUILDING RESEARCH ESTABLISHMENT, *Shrinkage of natural aggregates in concrete*, Digest No. 35, 2nd series, 1968, p. 5.

BUILDING RESEARCH ESTABLISHMENT, *Determination of chloride and cement content in hardened Portland cement concrete*, Information Sheet IS.13/77, July 1977, p. 4.

BUILDING RESEARCH ESTABLISHMENT, *Alkali-aggregate reactions in concrete*, Digest No. 258, Feb. 1982, p. 8.

CONCRETE SOCIETY, *Concrete core testing for strength*, Technical Report No. 11, 1976, p. 44.

CONCRETE SOCIETY, *Assessment of fire damaged concrete structures and repair by gunite*, Technical Report No. 15, 1978, p. 28.

CONCRETE SOCIETY, *Non-structural cracks in concrete*, Technical Report No. 22, 1982, p. 40.

GEE KIN CHOU, *Rebar Corrosion and Cathodic Protection—an Introduction*, published by Raychem Corp. California, p. 11.

HOBBS, D. W., *Drying Shrinkage and Movement of Reinforced Concrete*, Cement & Concrete Association, London, 42.522, April 1978, p. 19.

HOBBS, D. W., The expansion of concrete due to alkali–silica reaction. *Struct. Engr.*, Jan. 1984.

INSTITUTION OF STRUCTURAL ENGINEERS, *Appraisal of existing structures*, Report, July 1980, p. 60.

INSTITUTION OF STRUCTURAL ENGINEERS AND THE CONCRETE SOCIETY, *Fire Resistance of Concrete Structures*, published by the Inst. Struct. Eng., 1975, p. 59.

KOLEK, J., An appreciation of the Schmidt rebound hammer. *Mag. Conc. Res.*, **10**(28), March 1958, 27–36.

MCANOY, R., BROOMFIELD, J. P. AND DAS, C. S., *Cathodic protection—a long term solution to chloride induced corrosion?* Paper at International Conference, Structural Faults '85, Institution of Civil Engineers, London, April/May 1985, p. 7.

MALHOTRA, V. M., *Testing hardened concrete: non-destructive methods* (Monograph No. 9), Ref. M-9, 1976, ACI, p. 204.

NEVILLE, A. M., *Properties of Concrete*, 3rd ed., Pitman Books Ltd., London, p. 779.

PULLEN, D. A. W. AND CLAYTON, R. F., The radiography of Swaythling Bridge, *Atom*, Nov. 1981, No. 301, 3–8.

STEFFENS, R. J., *Structural Vibration and Damage*, Building Research Establishment, 1974.

STRATFULL, R. F., Half cell potentials and corrosion of steel in concrete. *Highway Research Record*, 1973, 12–21.

TOMSETT, H. N., The practical use of ultrasonic pulse velocity measurements in assessment of concrete strength, *Mag. Conc. Res.*, **32**(110), March 1980, 7–16.

VASSIE, P. R. W., *Evaluation of techniques for investigating the corrosion of steel in concrete*, TRRL Supplementary Report 397, 1977, p. 27.

Chapter 4

Non-Structural Repairs

A non-structural repair is one which, when completed, will not increase the load carrying capacity of the member nor of the structure. This type of repair should be carried out as a result of the investigations and diagnosis discussed in Chapter 3.

4.1 PREPARATORY WORK

No matter how carefully the investigation has been carried out it will not have revealed all the areas of defective concrete; in fact it was not intended to do this, but to establish the cause and general extent of the deterioration and provide the information needed to prepare a specification for the remedial work. Therefore, all parts of the concrete structure to which repairs will be carried out, must be 'hammer tested' to detect hollow and weak areas in addition to those areas where visible cracking, spalling and rust staining have occurred. The additional areas found by hammer testing should be clearly marked on the surface so that they will not be missed when the repair work gets under way.

All defective and deteriorated concrete must be removed, and rusted reinforcement must be cleaned of all corrosion products. This preparatory work is normally done by pneumatic tools which create considerable noise and vibration. In some cases this is unacceptable and then consideration has to be given to the use of high velocity water jets or flame treatment. Obviously, with the use of flame cutting equipment, great care must be exercised and specialist advice obtained from the suppliers of the equipment.

The use of high velocity water jets for cutting concrete was introduced into the UK in the early 1970s and although its use has increased slowly, the author feels that its potential has not been fully realised. The

FIG. 4.1. High velocity water jet cutting slot in reinforced concrete wall (courtesy F. A. Hughes & Co. Ltd).

technique has a great deal to recommend it for removing old and defective concrete, cutting concrete, exposing the coarse aggregate to provide a mechanical key for subsequent layers and for cleaning reinforcement of rust and corrosion products. The pressure at the nozzle is between $21 \, \text{N/mm}^2$ and $69 \, \text{N/mm}^2$. Comparatively little water is used, about 50 litres per jet per minute of which about one-third is dissipated as mist and spray. The jets leave the concrete and steel clean and damp. A very thin film of oxide will form quickly on the cleaned steel, but this does no harm. The pressure is determined by the work required to be done by the water jet; for cutting concrete, the pressure required to produce the necessary nozzle velocity would be about $41 \, \text{N/mm}^2$.

The author's experience is that provided the presence of water can be accepted, high velocity water offers the best chance of a quick clean job. The water will not cut the reinforcement and therefore this has to be cut separately.

However, there are occasions when the reinforcement must remain intact and therefore thermic lance and diamond drilling are not suitable.

FIG. 4.2. Close-up view of slot in Fig. 4.1 after cutting by high velocity water jet
(courtesy: F. A. Hughes & Co. Ltd).

An example of this is where a new concrete wall has to be joined and bonded to an existing one. Figure 4.1 shows high velocity water jets being used for this purpose, and Fig. 4.2 shows the concrete slot completed and ready to receive the new concrete. The wall was approximately 225–300 mm thick and 6·0 m high; the slot was tapered from 300 mm on the outer face to 50 mm at the rear. The work was completed in $7\frac{1}{2}$ h.

When it is necessary to cut through reinforcement the thermic lance can be used. This piece of equipment works on the principle that when carbon steels are heated at about 900°C (which is a bright red colour), they will burn in an oxygen-rich atmosphere; the flame arising from this combustion has a temperature of about 3500°C. This temperature is high enough to melt concrete, clay bricks and steel. The thermic lance can be used for a variety of work such as cutting holes in concrete, cutting up

FIG. 4.3. Spalled concrete beam being prepared for repair (courtesy: Cement Gun Co. Ltd).

FIG. 4.4. Precast concrete beam prepared for application of cement/SBR slurry bond coat, prior to applying repair mortar.

FIG. 4.5. Precast concrete column where chloride concentration was high, showing concrete removed in preparation for placing new concrete.

FIG. 4.6. Column shown in Fig. 4.5 after repair.

FIG. 4.7. Deteriorated column prepared for application of repair mortar and for fixing resistance probe to monitor future corrosion (courtesy: Gunac Ltd).

FIG. 4.8(a). Deterioration due to corrosion of rebars in r.c. column in high rise flats (courtesy: Gunac Ltd).

concrete sections and cutting openings in concrete walls. Under no circumstances should it be used where there is any risk of fire or explosion. For further details of this technique the reader is referred to the Bibliography at the end of this chapter.

Reinforcement can also be cleaned by wire brushing and grit blasting. The latter is more effective, but site conditions may prevent its use.

FIG. 4.8(b). Column in Fig. 4.8(a) after repair but prior to application of decorative sealing coat (courtesy: Gunac Ltd).

Having decided on the method or methods to be used for removal of the defective concrete, and the cleaning of the corroded reinforcement, it is necessary to consider, in many instances, how much concrete should be cut away. The author's experience is that in the majority of cases, significant corrosion is found to have taken place only on the outside half of the perimeter of the rebars (where the concrete cover is in some way defective). There may be some corrosion on the other half of the bars, but it should be remembered that it is accepted that some rust on rebars when the concrete is originally placed does no harm. The author can see no justification for cutting away good quality concrete and replacing it with mortar which may have a lower strength and higher permeability. Where corrosion has spread around the reinforcement, then the concrete should be cut away and the bar properly cleaned. Figures 4.3 to 4.8 show various stages in the repair of reinforced concrete structures.

4.2 EXECUTION OF REPAIRS

4.2.1 General Considerations

After the completion of the preparatory work and immediately before the application of the new mortar or concrete, the reinforcement and the surrounding concrete should be given a coat of Portland cement grout. The grout should be made of two parts of OPC to one part of a styrene–butadiene or acrylic latex, by weight. This should be well brushed into the concrete and applied evenly to the reinforcement. The object of the grout is to create an intense alkaline environment around the steel and to improve bond between the old concrete and the new mortar or new concrete. The grout should not be allowed to set before the mortar or concrete is placed in position; a maximum period of 20 min should be allowed between the two operations. The author is not in favour of the use of acid rust inhibitors, which are usually based on phosphoric acid. The use of acids in contact with concrete should be avoided. The material used for the repair has been described as 'mortar or concrete'. The reason for this is that both materials can be used successfully, but it is only practical to use concrete in positions where it can be properly compacted and there is sufficient thickness to accommodate the coarse aggregate. This is one of the reasons why most repairs are carried out with mortar.

Except for small repairs to chipped or broken arrisses the author favours the use of a modified Portland cement mortar, rather than a resin mortar. In this context, resin mortar means a mixture of an organic polymer and fine aggregate, in other words the cement is replaced by the polymer. These resin mortars have their uses for special purposes, but are very expensive compared with the same volume of cement/sand mortar, and for general repairs to building structures the extra cost is not justified.

When considering the use of resin based mortars attention should be given to the fact that the coefficient of thermal expansion of these resin mortars is likely to be in the range of $25–35 \times 10^{-6}$, compared with cement based mortar and concrete of $7–13 \times 10^{-6}$. A 'pure' epoxy, i.e. the resin without any filler, may have a coefficient of thermal expansion of about 60×10^{-6}.

The need to obtain a reasonable compatibility of strength between the base concrete and the repair mortar/concrete is very important. This can create problems when the base concrete is rather weak, as this low strength is often accompanied by higher than desirable permeability. If the repair mortar is much stronger than the base concrete there will be a tendency for debonding to occur. The bond at the interface can only be as strong as the weaker of the two materials which are bonded together. In some cases it may be necessary to use a light galvanised or stainless steel mesh pinned to the base concrete.

This need for a reasonable compatibility between the base concrete and the new repair mortar/concrete can sometimes result in having to use a mortar or concrete with a lower cement content then is desirable for the adequate protection of the reinforcement. In such a case the provision of a high quality protective coating to the concrete is essential.

Three of the most important requirements in a satisfactory repair are:

(a) To ensure as far as practicable that all defective concrete, and rust on reinforcement, has been removed.
(b) To ensure the best possible bond between the old and the new work.
(c) To ensure that the new mortar or concrete is as impermeable as possible, compatible with the selected mix proportions.

An additional factor of considerable practical importance is that it is invariably found that there are many areas where the cover is inadequate but the reinforcement has not corroded to any significant extent and there is no cracking. Also, it must be remembered that in most repaired areas, the thickness of the repair mortar will not provide what is now considered adequate cover, because it has to be finished in line and level with the adjoining concrete. The repair mortar merely provides the same depth of cover that was originally provided by the concrete when the structure was erected. It is of course hoped and expected that the new mortar will be more impermeable than the original concrete, but the thickness is governed by the surface features of the adjacent concrete.

Reference has been made in Section 4.1 of the need to ensure that the repair material (mortar or concrete) is 'compatible' with the base concrete to which it will be required to bond. The important factors for compatibility include compressive strength (with cement based mortar/concrete), the cement content and water/cement ratio.

The compressive strength should be as compatible as practicable with that of the base concrete; it is bad practice to try to bond a high strength mortar to a comparatively weak substrate. The cement content and the water/cement ratio should be kept as low as possible, commensurate with adequate workability for proper compaction. The great majority of repairs are carried out with mortar, and for this, a clean sharp sand should be used and the selected mix proportions should be based on weight rather than volume. The use of weigh-batched, prepacked sand is desirable as this will help considerably to ensure uniformity of quality of the mortar. In the opinion of the author, exact compliance with the recommendations for grading in relevant British Standards is not essential. A concreting sand to Table 5, Grading M, BS 882:1983 is suitable, and the same applies to a sand from Table 1, Type A, BS 1199:1976, Amendment No. 3. Where there is a choice, the use of the coarser grade is preferred because finer sand increases the water demand of the mix.

The author favours the use of a styrene–butadiene rubber emulsion in the mortar mix; the concentration of SBR would be 10 litres to 50 kg cement. SBRs which have been manufactured for use in the construction industry are quite compatible with both Portland and high alumina cements. The SBR reduces permeability and shrinkage, and increases bond and resistance to chemical attack while reducing the rate of carbonation.

Having decided on the materials and the mix proportions the next question is the method of application of the mortar. The author is of the opinion that the most effective method is by spray as this helps to ensure complete contact between the mortar and the whole of the substrate including the reinforcement, to which it is applied, and it ensures better bond than with a hand applied mortar. However, with very small areas of repair, hand application is the only practical way to do the job. With both methods, the final finish must be done by hand trowelling and conscientious work is essential for a satisfactory result, that is, a dense, well compacted mortar of low permeability.

It is essential that all practical measures should be taken to ensure proper curing of the newly placed mortar or concrete. This can present some problems in those cases where the repaired areas are small and widely spaced. Even with reasonable attempts at curing, it is often found that fine hairline cracks develop around the perimeter of the repaired areas due to drying shrinkage. Unless steps are taken to seal these cracks, they are likely to form the nucleus of fresh deterioration as they will tend to widen due to frost action. This problem is dealt with in the next Section.

It is therefore recommended that where repairs of any magnitude are carried out, the whole surface of both the repaired areas and the areas which have remained without repair should be sealed with a durable coating. This should reduce the effects of low cover and permeable concrete and seal any fine shrinkage cracks which may occur in the new work.

The ideal characteristics of such a coating would include:

(a) Excellent bond to the substrate.
(b) Durable, with a useful life exceeding 20 years.
(c) Minimum colour change and little 'chalking'.
(d) Maximum permeability to the passage of water vapour from the concrete substrate.
(e) Minimum permeability to the passage of external moisture (rain, dew, etc.) through to the concrete.
(f) Low permeability coefficient to the passage of oxygen and carbon dioxide from the air to the concrete.
(g) To be available in a reasonable range of attractive colours.

The author does not know of any coating on the market which fulfils all the above conditions; he has found it very difficult to obtain clear and factual information from suppliers on factors (a)–(f) above.

The Building Research Establishment at Garston have done a considerable amount of research into the characteristics of a wide range of coating materials. This is summarised in Digests Nos. 197 and 198. Reference to Table 1 in Digest 197 (1982 edition) shows that there is no material which possesses all the desirable properties. It is therefore necessary for the Engineer specifying the 'decorative sealing treatment' to compromise and select the basic material according to his assessment of site conditions and tests on the material. To do this he will need information from the manufacturers and discussions with them. In coming to this compromise, the author generally favours low permeability to external moisture in preference to high permeability to vapour. As new materials backed by enthusiastic claims by the manufacturers are constantly coming onto the market the author is unable to make any specific recommendations. Reference can usefully be made to BS 3900—Tests for Paints, Part E6—Cross-cut Test and Part E10—Test for Adhesion. Recommendations for the implementation of these tests on coatings for marine structures are given in Chapter 10, Section 10.3, where the use of a coating which can serve as a curing membrane as well as a long term durable surface sealant is discussed. There are obvious advantages if the new mortar repairs can be properly cured with a primer which forms the first coat in a two-coat application of a decorative/sealing coat but this can only be done in special circumstances.

4.2.2 The Repair of Honeycombed Concrete

The previous sections have dealt with the principles of general repairs to concrete members where the defects are usually only too obvious. However, cases occur where the defects are invisible, such as inadequate compaction and honeycombing of the concrete within a member, such as a beam, column or wall. There may be some signs on the surface which to an experienced eye suggest that all may not be well in the interior of the member. Factors which are likely to cause this type of trouble include congested reinforcement, low slump concrete, breakdown in poker vibrators, premature stiffening of the concrete as it is being placed, movement of the formwork.

The suspected concrete should be checked by cutting out, or by an ultrasonic pulse velocity survey (upv), or gamma radiography, or by taking cores. If this investigation demonstrates that the concrete is defective then it either has to be repaired or removed; this latter course may involve demolition of the member. Demolition is difficult and expensive and can cause serious delay in the completion of the structure, and in fact is seldom justified. The removal of the honeycombed or under-compacted concrete and its replacement by new concrete is usually a satisfactory solution. The major difficulty in such work is to ensure that the new concrete is itself fully compacted. The author has found that the procedure described below, when carefully carried out, will give satisfactory results. This refers to the repair of a load bearing wall or column where the defects are near the base, but the principles can be adapted for other members.

The concrete must be cut away to expose the under-compacted and honeycombed area. It may be prudent to provide temporary support to the member while the repair is in progress.

When the extent of the honeycombing has been determined by such work, proceed to remove all substandard concrete. The cutting away can be carried out by percussion tools, thermic lance, or high velocity water jets, whichever is more appropriate. Percussion tools produce noise and

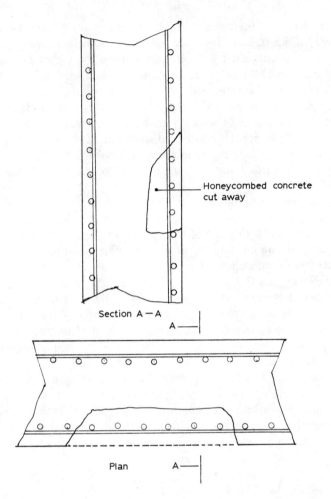

Honeycombed concrete cut away

Section A—A

A

Plan A

FIG. 4.9. Method of repair of honeycombed concrete in wall or column.

vibration. Thermic lances may damage the reinforcement, but do not produce noise and vibration. Information on the use of high velocity water jets has already been given. Whichever method is used, sufficient concrete must be cut away to provide adequate space for placing and compacting the new concrete (see Fig. 4.9).

Even under the most favourable conditions, compaction of the concrete is likely to be difficult, and unless all precautions and great care are taken, it may well be impossible. The essentials are a cement-rich mix, well graded aggregates, low w/c ratio and sufficient workability for proper compaction. This will certainly require the use of a plasticiser. Consideration should be given to the use of a superplasticising admixture. Some brief information on these materials is given in Chapter 1.

It is not suggested that the above will necessarily result in the repaired member having the same strength as the original was intended to have. However, if allowance is made for redistribution of stress between the old and new concrete and the reinforcement, and provided the work has been properly carried out, the small reduction in strength should be insignificant.

For honeycombing in non-load bearing members, pressure grouting with a cement grout, executed by a specialist firm, should be adequate. Even so, because of the cost of the equipment needed for this work, for small jobs, cutting out and repair may be more economic.

Apart from compaction, there is the problem of ensuring a weather tight joint between the old and new concrete; this assumes even greater importance if the cut-out section penetrates right through the wall. A clean, dust free surface to the old concrete is essential. While a brush coat of grout on the surface of the old concrete, applied immediately before the new concrete is placed, will help ensure good bond, it is essential that the concrete placing follows immediately behind the grout. If this cannot be ensured then it is better to omit the grout. If a grout is used, it is recommended that it consists of 2 parts ordinary Portland cement to 1 part styrene–butadiene latex emulsion by weight. If 10 mm aggregate is used, then the sand content should be increased to about 45% and there should be an increase in the cement content. The new concrete should be cured in the usual way for at least four days.

Repair with Cement/Sand Mortar.

For small areas, particularly those of shallow depth, the use of a cement/sand mortar is a practical and satisfactory solution. The honeycombed concrete is cut out as previously described above, and the void is carefully filled with the mortar. The mortar should have as low a w/c ratio as possible so as to reduce drying shrinkage. The actual consistency of the mortar will depend on the exact condition of placing. In some cases, a very dry mix (earth dry) can be used and compacted into the void with hand or pneumatic tools. Honeycombing at a kicker joint at the base of a wall can often be repaired in this way.

For normal hand applied mortar, the mix proportions are usually about 1 part OPC or RHPC to 3 parts well graded sand. The author recommends the use of a styrene–butadiene based latex emulsion as a gauging liquid, with water added only to obtain the necessary workability. The use of the latex will improve bond with the base concrete, and help to reduce permeability and shrinkage.

Effective and practical measures should be taken to cure the mortar in the repaired areas, but if an SBR latex emulsion is used, the start of curing is usually delayed by 12–24 h, or as directed by the suppliers. For larger areas, particularly those of relatively shallow depth, say down to 100 mm, pneumatically applied mortar is likely to be a better solution than hand applied mortar. The reason for this is that the mortar will be better compacted and therefore denser and more impermeable. Gunite, in which the mortar leaves the nozzle of the 'gun' at high velocity, is a structural material in its own right. The usual mix proportions are 1 part OPC to 3 parts well-graded concreting sand, with a w/c of about 0·35. The resulting mortar can have a compressive strength in excess of $40 \, N/mm^2$. It is of course only practical to use gunite where there is sufficient repair work to justify bringing the equipment to the site.

With hand applied mortar, and to a lesser extent with gunite, a fine hair crack is liable to appear around the perimeter of the repaired area. To seal this, it is recommended to brush the new mortar with a stiff bristle brush, and the surrounding concrete with a wire brush, to remove any relatively friable surface layer. This treatment should extend for 150 mm each side of the boundary between the old and new work, and should be carried out as late in the construction process as possible. As soon as this preparation is completed, a heavy brush coat of grout should be applied and worked well into the surface. The grout should be composed of 2 parts OPC and 1 part styrene–butadiene latex emulsion by weight.

Surface Sealing of Honeycombed Concrete
In some cases it will be decided that cutting out the concrete is not necessary nor desirable, and surface treatment of the area might then be adopted as an alternative to pressure grouting or in addition to pressure grouting.

The application of a resin coating of suitable formulation is only practical where the surface of the concrete is reasonably sound and the honeycombing is below the surface. Sometimes interior honeycombing shows itself by a few surface defects and it is only when an ultrasonic pulse velocity survey is carried out or cores are taken, that the true extent of the honeycombing is revealed. °

The author feels that the strongest argument against a coating is that sooner or later it will have to be renewed. However, this also applies to protective treatment of ferrous metals and to the sealants used in joints. The great advantage of this method is that all parts of the process (surface preparation and application of the coating) are visible and open to inspection, whereas pressure grouting cannot be seen and is something of a hit or miss business when used on its own. The surface of the concrete should be prepared by removal of the laitance and light exposure of the coarse aggregate. An exposure of 3 mm is the maximum which is needed, and 1 mm is usually sufficient.

It is generally advisable for the resin to be formulated to bond to damp concrete, and of course it must possess adequate durability under the operating conditions of the structure. Epoxide resins and poly-urethanes, and sometimes combinations of both, are the resins in general use.

It is advisable to use a low viscosity primer, followed by successive coats of a high build resin, to give a total thickness of not less than 0·75 mm. The treated area should extend well beyond the estimated boundary of the honeycombing, i.e. by not less than 300 mm.

For drinking water reservoirs all repair materials used for the repair work should be non-toxic, non-tainting and should not support bacterial growth.

Pressure Grouting of Honeycombed Concrete
A cure can often be effected by pressure grouting. A description of pressure grouting to seal leaks in basement walls and floors is given in Advisory Leaflet No. 52, issued by the Department of the Environment. The impression given by the Advisory Leaflet is that pressure grouting is relatively straightforward and that if the procedure described is followed, the leaks will be sealed. Unfortunately the author's experience has caused him to be rather pessimistic about the result of pressure grouting on its own unless it is carried out by an experienced firm. Even then, disappointments are not uncommon. It is relevant to note that specialist firms will very seldom guarantee success and insist on working on a time-plus-materials basis. This means that they will not give a fixed price nor will they guarantee complete watertightness on completion.

The grout used is usually based on Portland cement with an admixture, which is often PFA. Sometimes non-shrink grouts are used. These are often premixed powders to which water is added. Another method is to use a special admixture in the mixing water, which the manufacturers claim will reduce or even eliminate shrinkage.

In the author's opinion, it is best to combine pressure grouting with surface sealing. The surface sealing in existing (not new) structures can usually be applied only to one face of the defective concrete.

4.3 THE REPAIR OF BINS AND HOPPERS HOLDING COARSE-GRAINED MATERIALS

Two materials in general use in industry and which are stored in open bins and hoppers, are limestone and coke. Both of these can cause severe abrasion to the inside of the structure but limestone has no adverse effect on the concrete. On the other hand, coke contains chemical compounds and when it is slaked with water the resulting effluent may be very aggressive to Portland cement, due principally to the presence of mineral acids and sulphates in solution.

The deterioration of concrete in these structures usually follows a particular pattern. Inside there is abrasion, particularly on the surface of the sides and bottom of the hopper, and in the case of coke holding bins, there is likely to be chemical attack as well. With older structures, rusting of reinforcement, cracking and spalling of the concrete may occur both on the inside and externally. This is largely due to rather poor quality concrete, and inadequate cover to the reinforcement. To this is sometimes added structural distress resulting from underestimation in the original design of the actual stresses developed during filling and emptying.

The logical, and best material for the repair of these structures is high quality reinforced gunite. High quality gunite is dense and impermeable and resistant to abrasion. These are the qualities in mortar and concrete which form the first line of defence against chemical attack.

The use of sulphate-resisting Portland cement is generally effective in preventing deterioration of the concrete by sulphates in solution (except ammonium sulphate), up to a concentration of about 5000 ppm. Above this figure, and where acids are present, consideration should be given to the use of high alumina cement. One of the basic requirements in the satisfactory performance of HAC concrete and mortar is a low initial w/c ratio. This is met by high quality gunite which has a w/c of about 0·35.

Further information on HAC is given in Chapter 1. Abrasion resistance of cementitious materials is directly related to compressive strength. The 28 day equivalent cube strength of good quality gunite should be in the range of 40–50 N/mm².

Sometimes the slope provided for the inside of the hopper bottom is insufficient and this causes trouble when emptying the bin. A slope of something over 50° may be necessary, and to provide this new slope in concrete requires the use of a top shutter. However, this brings problems of placing, compaction and finish. As with all work involving bonded toppings, careful preparation of the base concrete is essential.

4.4 REPAIRS TO CAST AND RECONSTITUTED STONE

Cast stone and reconstituted stone are both a type of precast concrete; BS 2847— Glossary of Terms for Stone Used in Building, defines cast and reconstituted stone as:

'A building material manufactured from cement and natural aggregates for use in a manner similar to and for the same purpose as natural building stone.'

Cast stone is covered by BS 1217 and another relevant British Standard is BS 5642, Parts 1 and 2—Sills and Copings, which includes precast concrete and cast stone.

The requirements in BS 5642 with regard to concrete quality and cover to reinforcement are rather more stringent than in BS 1217. The requirements in BS 5642 are similar to those of BS 8110, for grade 30 concrete, and for the cover to reinforcement.

Unfortunately, many copings, sills and mullions in cast/reconstructed stone have, in the past, proved to be unsatisfactory from the point of view of durability due to the corrosion of reinforcement. Except in the case of mullions, the reinforcement is nominal and is there for handling purposes. As the units are generally of small cross section, the corrosion of the reinforcement can result in virtual destruction of the units. For this reason, it is in many cases better to renew the units completely.

If it is decided to repair the units, then special care is needed in the selection of the mix proportions and the type of aggregate. It is usually impossible to obtain a good match between the colour of the existing units and the new repair mortar. The provision of a decorative sealing coat is then essential because sills, copings and mullions are architectural features of the building.

FIG. 4.10. View of deteriorated cast stone window sill.

Due to the slim cross section of the units and the usual rather low strength of the concrete, it is often necessary to use a stainless steel mesh pinned into the units and embedded in the new mortar. The use of an SBR bond coat and the inclusion of the SBR in the repair mortar is recommended. Figure 4.10 shows a deteriorated cast stone sill.

4.5 THE WEATHERING OF CONCRETE BUILDING STRUCTURES

The effect of weathering on a concrete structure depends on the type and quality of the concrete and the degree of exposure. In the vast majority of cases, repairs to the structure are not required as a result of weathering alone. In fact in country districts weathering can result in a marked improvement in the appearance of the concrete without causing any deterioration at all.

Weathering may be defined as changes in the surface layers and particularly in surface appearance, brought about by the weather. The author would not include the carbonation of the concrete in the general

FIG. 4.11. Weather staining of precast concrete units on office block in London
(courtesy: Cement & Concrete Association).

definition of weathering. Some basic information on carbonation has
been given in Chapter 2.

The causes of weathering are numerous and include the following:

    (a) Dirt in the atmosphere deposited on the surface of the concrete by
        wind, rain and fog.
    (b) Deposits washed onto the surface from adjoining surfaces.
    (c) Chemical attack from aggressive compounds in the atmosphere.
    (d) Fungal and algal growths on the surface.
    (e) The general effect of wind, rain, and atmospheric temperature
        changes.

The deposits formed under (a), (b) and (d) may in the long term provide
a protective coating against attack from (c).

Concrete surfaces, particularly plain ones, are very vulnerable to
weather staining. This is due to a number of factors, including the fact
that the surface of concrete is absorbent and that it is often cast in
comparatively large light coloured areas. Brick and stone are also
absorbent (more so than concrete), but the units are very much smaller

FIG. 4.12. Weather staining and water percolation on bridge abutment (courtesy: Cement & Concrete Association).

and each small area is separated from adjoining ones by joints made of different material and having a different texture and colour tone.

The inherent problems associated with the weather staining of plain concrete are the main reason for the move away in recent years from this type of surface, to that of exposed aggregate and moulded finishes. Figures 4.11 and 4.12 shows weather staining of concrete on a new office building in London. Weather staining in itself is not detrimental to concrete and it is only when there is chemical attack from atmospheric pollution that damage occurs. As stated above, a thick layer of grime is likely to help prevent or reduce attack. However, in a chemically aggressive environment, as exists in some industrial towns, when the building is cleaned, the newly exposed surface of the concrete or cast stone will be exposed to intensified attack.

Smoke pollution has been substantially reduced during the past 25 years in many of the large cities in the UK. In London this reduction is particularly noticeable and it is therefore all the more unfortunate that the level of sulphur dioxide (SO_2) has not decreased and is now probably higher than in any other town in the country. When sulphur dioxide is dissolved in rain water or fog, sulphurous and sulphuric acids are formed which are very aggressive to Portland cement concrete and calcareous materials such as limestone.

It is therefore advisable to give these facts careful consideration before embarking on a scheme of external cleaning. The cleaning will certainly improve the appearance of the concrete but may result in accelerated deterioration, although this is likely to apply more to certain types of stone masonry than to Portland cement concrete. Where it is feared that deterioration may occur, it is difficult to put forward recommendations which will not alter, to some extent, the light reflecting properties of the surface. Also, all applied coatings undergo some degree of colour change. Ideally, this type of coating should be colourless, of low viscosity, very durable, bond well to the substrate, and should not lose its transparency. The author knows of no material which fulfils all these conditions entirely satisfactorily, and so some compromise is required.

Silicones probably give the best results from the point of view of appearance, but are short-lived. Orthosilicates have an appreciably longer life, but are rather more visible than silicones. Low viscosity epoxide resins are very durable but are likely to change the appearance of the concrete or cast stone.

However, the application of waterproofing liquids should be carried out with care, particularly if it is not intended to treat complete wall areas. This is referred to again in more detail later in this chapter under the waterproofing of walls.

A new type of material has come onto the market in recent years, known as silanes, which are claimed, with some justification, to be appreciably superior to silicones for reducing water permeability of many types of building materials including concrete. They possess great penetrating powder and increase surface tension in the fluid water and thus greatly reduce absorption of the treated materials. The estimated 'life' is in excess of ten years.

There is another type of weathering which can cause deterioration of concrete and cast stone, and this is frost action. This is unlikely to occur in the UK if good quality concrete and cast stone are used, but poor quality materials can be damaged in this way on exposed sites. With porous material, such as poorly compacted concrete, and concrete and cast stone made with a high w/c ratio, moisture penetrates the surface layers and when it freezes it expands and the expansion causes spalling. It should be particularly noted that high quality concrete is much more impermeable, and because of its higher strength is better able to resist the stresses caused by the expansion of absorbed moisture when this freezes. On very exposed sites, it is advisable to use air-entrained concrete. This type of concrete is now recommended for all external paving as it will successfully resist the disruptive effects of de-icing salts. Air-entrained concrete and mortar should not be confused with aerated concrete and mortar which are quite different materials. Brief information on air-entraining agents has been given in Chapter 1.

The air-entraining resin used must be carefully dispensed into the mixing water so as to ensure uniform distribution throughout the batch of concrete or mortar. The dosage must be such that the total air content of the mixed material is in the range of $4\frac{1}{2}\% \pm 1\frac{1}{2}\%$. The entrainment of this amount of air has the effect of reducing the compressive strength of the concrete by about 10%, but at the same time it acts as a workability aid and thus the w/c ratio can be reduced, which in turn will increase the strength of the concrete. In repair work, impermeability and durability are usually of greater importance than strength and so the small net reduction in strength is not important. Reference should be made to BS 6270, Part 2—Cleaning and surface repair of buildings—Concrete.

4.6 CATHODIC PROTECTION OF STEEL IN CONCRETE STRUCTURES

4.6.1 General Principles of Cathodic Protection

In above-ground steel structures the usual method of protecting steel against corrosion is by means of coatings. These may be of another metal, as in galvanising or metal spraying, or various paint systems, as well as the more recently introduced use of plastic films.

In Chapter 2 it was stated the corrosion of metals in a conducting environment is electrochemical in nature, and this is due to all metals attempting to revert to their natural ore state. Exposed steel in a moist environment will corrode due to differences in electrical potential on the surface of the metal itself. These areas form anodes and cathodes, which, by definition, permit a flow of electric current from anode to cathode. It should be noted that electron flow, which is by inference a part of the same process, is in the reverse direction. The multitude and extent of the consequent electrochemical reactions result in the generation of a continuous electric current made up of a multiple of small anodic and cathodic areas. With steel and some other alloy structures buried underground or immersed in water (e.g. marine structures) corrosion control is often exercised by electrical methods and known as 'cathodic protection'. This is usually achieved by the application of a continuous direct current from an external source. It is not possible, nor desirable in this book, to enter the detailed theory and practice of cathodic protection, so only brief mention will be made of the principles involved.

When two dissimilar metals are joined together in an electrolyte a current is produced as a function of the electrochemical series of metals. For example, if steel and zinc are connected, a current will flow from the zinc to the steel because the zinc is anodic to steel. This reaction takes place only if there is an electrolyte present between the two dissimilar metals, although it does not necessarily have to be water, inasmuch as severe corrosion can take place in soils ranging from wet clay to rock. With the zinc/steel combination just mentioned, the zinc will corrode and the steel will not, and the overall effect is for the steel to have become inhibited against corrosion attack by virtue of the anode material sacrificing itself to afford protection of the steel. Hence the term 'sacrificial anode protection'.

Alternatively, the same result can be obtained by the introduction of a controlled d.c. electricity application to the structure from either an a.c. supply source, diesel generators, thermo-electric equipment, solar cells or even high efficiency windmills. The structure is connected to the negative supply (or cathode) and the positive to an introduced anode which is chosen to have semi-inert (or non-corrodable) properties. This is known as the 'impressed current' method.

Thus there are two practical ways of introducing cathodic protection which can be summed up as:

(a) By connecting the steel to a metal which is less noble in the electrochemical series. The following list shows relative activity of metals which varies from the highly reactive alkaline metals such as potassium and sodium to noble metals like gold. The list indicates a basic series in which the metals high on the list will become anodic to those lower down and hence provide protection. This is the basis of the principle previously described as sacrificial anode protection, and depends on one metal being designed to corrode and so prevent another corroding.

Sodium
Magnesium
Zinc
Galvanised iron
Aluminium
Mild steel
Cast iron
Stainless steel (Cr based)
Lead
Bronze
Brass
Copper
Stainless steel (passive)
Silver
Gold

Note the relative positions of commonly used metals or alloys, e.g. galvanised iron/steel/brass/copper/stainless steel, which is the basis of most forms of introduced electrolyte corrosion damage.

(b) By the application of an external current of sufficient intensity to 'swamp' the corrosion current. The impressed current is d.c. with the positive side of the output connected to a purpose made anode, and the negative to the structure being protected. The anode can, in theory, be of any material which will conduct electricity, but in practice the range of suitable materials is limited principally to ensure system longevity. Anodes in this system are comparatively inert and are designed to last much longer than sacrificial anodes. When correctly selected and used in a designed system, such anodes must have a useful life of 20 years and more. The materials available are based on graphite, high silicon-iron, platinised titanium, tantalum or niobium or the lead/silver alloys which are principally used in marine installations. Although platinum and its alloys appear expensive they are becoming more popular because of their electrochemical properties and light weight.

In considering the application of cathodic protection to a structure, the following questions arise and must be satisfactorily answered:

(i)   Surface area to be covered and its treatment, e.g. bare, old paint, newly applied high grade coating.

(ii) The environment surrounding the structure, e.g. wet conditions, pH, chemical characteristics.

(iii) What type of protection might be the most suitable in the given circumstances, i.e. sacrificial anodes or impressed current.

(iv) The general circuit configuration, including the number, location and type of anodes to be used.

(v)   Is there any likelihood of the protective current flowing outside the network which is supposedly being protected? If this risk is likely to arise, then anode locations must be changed or interference testing contemplated.

The most practical way of assessing any form of electrolyte, whether liquid or solid, is by testing its electrical resistivity utilising special megger type equipment. The following table indicates in a very general way the relationship between the environment and its corrosiveness in terms of resistivity:

TABLE 4.1
(*from B.S.I. CP 1021:1973—Cathodic protection*)

Resistivity (ohm. metres)	Degree of corrosiveness
Up to 10 ohm.m	Severely corrosive
10 ohm.m—100 ohm.m	Moderately corrosive
100 ohm.m and above	Slightly corrosive

It will be seen that virtually no environment can be considered as non-corrosive, because every metal has a tendency to revert back into the basic form from which it was extracted as an ore. The environment controls the speed of this reaction, and ranges between highly corrosive conditions, such as found in sea water to virtually non-corrosive conditions in granite rock.

It is generally desirable that buried or immersed steelwork should, in addition to cathodic protection, be complemented with suitable coatings or wrappings. This additional protection is provided in order to reduce the rate of dissolution of the sacrificial anodes, and so prolong their useful life, or if impressed current is utilised, to reduce the power consumption.

4.6.2 General Considerations for the Protection of Reinforcement in Existing Structures

To stop the corrosion of rebars in an existing structure where the concrete contains a high concentration of chlorides (say, for example over 1% by weight of cement), requires:

(a) The provision of a complete inert barrier around each rebar over its full length, or
(b) the installation of an effective system of cathodic protection.

Method (a) is only likely to be possible in theory, except in very exceptional circumstances e.g. complete removal of all concrete around rebars.

For method (b) the cathodic protection system should be designed and installed so as to eliminate differences in potential between different parts of the reinforcing cage in each member to be protected.

Unfortunately, concrete has a high resistance to the flow of electric current and the cathodic protection systems which are effective for steel pipelines and steel tanks and similar structures did not give satisfactory results for reinforced concrete structures unless they were designed and installed when the structure was built. This applies to the use of both sacrificial anodes and impressed current. In *Concrete Structures: Repair, Waterproofing and Protection* (Perkins, 1976), the author included a short section on cathodic protection in the chapter dealing with the repair of concrete marine structures. He expressed the view that the use of cathodic protection in preventing corrosion of rebars in existing concrete structures held considerable promise provided basic difficulties could be overcome.

The cathodic protection of corroded prestressing wires in prestressed concrete pipes was successfully carried out in Israel in the late 1950s and was reported by Unz, Spector and others. It is understood that the corrosion was due to chlorides in the sand used for guniting in the prestressing wire. There is a danger that with an impressed current, hydrogen may be evolved and this could result in hydrogen embrittlement in highly stressed steel wires.

Development and research work by Spellman and Stratfull in California, and Manning in Canada in the 1960s and early 1970s on the use of cathodic protection in salt damaged concrete bridge decks is well known. A detailed Report on the cathodic protection of bridge decks in Ontario by Fromm and Wilson was issued by the Ministry of Transportation and Communications, Ontario, in mid-1970. The conclusions of this Report were:

(a) Cathodic protection of bridge decks is feasible.
(b) Both slab and post-tensioned voided decks can be protected by this method.
(c) The cost of the complete repair, including the cathodic protection, is small compared to the cost of deck replacement. The power required for protection is very small.

The main difficulties encountered in providing cathodic protection to rebars in an existing structure have been largely overcome by the introduction of effective continuous anode techniques, of which there are a number (patented or with patents pending). The two of which the author has some knowledge are the conductive coating system developed by Taywood Engineering, and the Ferex anode mesh developed by

FIG. 4.13. Cathodic protection of concrete column by conductive coating and impressed current (courtesy: Taywood Engineering).

FIG. 4.14. View of Ferex anodes installed on reinforced concrete beams and columns, prior to guniting (courtesy: Raychem Ltd).

Raychem Corporation. References on these two systems are given in the Bibliography at the end of this Chapter.

Very briefly the Taywood system uses a special conductive coating on the surface of the concrete which forms an external anode; the passage of an impressed D.C. current forces the reinforcement in the concrete to become cathodic, thereby inhibiting corrosion. Figure 4.13 shows such a system installed on external reinforced concrete columns which contained a high concentration of chlorides arising from calcium chloride added during the manufacture of the units. The basic conductive coating is black, but it can be given a decorative coating in a limited number of pastel shades, and is thus particularly suitable for building structures.

The Raychem impressed current system using Ferex anode mesh is shown in Fig. 4.14. This consists of an open weave mesh of flexible polymeric anode strands which is intended to ensure uniform current distribution as each rebar is not far from an anode strand. The mesh is covered with concrete or gunite.

BIBLIOGRAPHY

AMERICAN CONCRETE INSTITUTE, *Concrete repair and restoration*, ACI, Compilation No. 5; ref. C–5, 1980, p. 118.

BARNHART, R. A., Memorandum, FHWA position on cathodic protection systems, Federal Highway Administration, US, April 1982.

BRITISH STANDARDS INSTITUTION, *Building papers (breather type)*, BS 4016:1972.

BRITISH STANDARDS INSTITUTION. *Methods of test for inorganic thermal insulating materials*, BS 2972:1975.

BRITISH STANDARDS INSTITUTION, *The structural use of concrete*, BS 8110:1985, Parts 1 and 2 (replaces CP 110).

BRITISH STANDARDS INSTITUTION, *Code of Practice for the painting of buildings*, BS 6150.

BRITISH STANDARDS INSTITUTION, *Methods of test for paints*, Groups A to H, in 57 Parts.

BRITISH STANDARDS INSTITUTION, *Code of Practice for the protective coating of iron and steel structures against corrosion*, BS 5493.

BRITISH STANDARDS INSTITUTION, *Cleaning and surface repair of buildings*, Part 2: Concrete, BS 6270.

BRITISH STANDARDS INSTITUTION, *Cathodic protection*, CP 1021.

BUILDING RESEARCH ESTABLISHMENT, *Painting walls*, Parts 1 and 2, Digests 197 and 198, 1977 and 1982.

BUILDING RESEARCH ESTABLISHMENT, *The durability of steel in concrete*, Parts 1, 2 and 3, Digests 263, 264 and 265, July to Sept. 1982.

CIRIA, Report No. 93, *Painting steelwork*, 1982, p. 124.

CONCRETE SOCIETY, *Guide to precast concrete cladding*, Technical Report No. 14, 1977, p. 24.

CONCRETE SOCIETY, *Durability of tendons in prestressed concrete*, Technical Report No. 21, 1982, p. 8.

CONCRETE SOCIETY, *Non-structural cracks in concrete*, Technical Report No. 22, 1982, p. 40.

CONCRETE SOCIETY, *Repair of concrete damaged by reinforcement corrosion*, Technical Report No. 26, Oct. 1984, p. 31.

GEE KIN CHOU, *Rebar corrosion and cathodic protection*—an introduction, Raychem Corp., California, p. 11.

HAUSMANN, D. A., Criteria for the cathodic protection of steel in concrete structures, *Materials Protection*, 8(10), Oct. 1969, 23–25.

HAWES, F., *The Weathering of Concrete Buildings*, Cembureau, Cement & Concrete Association, 1981, p. 109.

HIGGINS, D. D., *Efflorescence on Concrete*, Cement & Concrete Association publication 47.104, 1982, p. 8.

MCANOY, R., BROOMFIELD, J. P. AND DAS, C. S., *Cathodic protection—a long term solution to chloride induced corrosion?* Paper at International Conference, Structural Faults '85, Institution of Civil Engineers, London, April/May 1985, p. 7.

MANNING, G. D. AND RYELL, J., *Durable bridge decks*, Ministry of Transportation & Communications, Ontario, Report RR.203, April 1976.

MATHER, K., *Concrete weathering at Treat Island, Maine*, ACI, International Conference, Canada, August 1980, pp. 1–1–111.

NATIONAL ASSOCIATION OF CORROSION ENGINEERS, *Solving rebar corrosion problems in concrete*, 15 Papers, NACE Seminar, Chicago, 1982.

NEW ZEALAND PORTLAND CEMENT ASSOCIATION, *Weathering of Concrete Buildings*, July 1981, p. 5.

PERKINS, P. H., *Concrete structures: Repair, waterproofing and protection*, Applied Science Publishers, London, 1976.

SPELLMAN, D. L. AND STRATFULL, R. F., Chlorides and bridge deck deterioration, *Highway Research Record*, No. 328, Transportation Research Board, Washington D.C., 1970, pp. 38–49.

The Journal Concrete, *Concrete repairs*; a selection of fourteen articles reprinted from *Concrete*, 1984, p. 60.

UNZ, M., Cathodic protection of prestressed concrete pipe, *Corrosion*, **16**, June 1960.

Chapter 5

The Structural Repair of Reinforced Concrete

5.1 GENERAL CONSIDERATIONS

Errors in design and the overloading of part of a structure which result in structural failure and collapse are fortunately very rare. When this does happen, a great deal of publicity (some of which can be very ill-informed), is given to the incident.

A natural question at this stage is, what is a failure? This is difficult to answer because it is often a matter of degree, and the importance attached to the defect. A failure can be considered as occurring in a component when that component can no longer be relied upon to fulfil its principal functions. For example, deflexion of a floor which caused a certain amount of cracking in partitions, could quite reasonably be considered as a defect, but not a failure; while excessive deflexion resulting in serious damage to the partitions, ceiling and floor finish would be classified as a failure. An error in design may, fortunately, come to light before or during construction, without any harm or damage being caused.

Structural distress in a new unused building, showing as tensile and/or shear cracks of unacceptable width, bowing of vertical members such as columns and walls, and excessive deflexion in horizontal members, indicates that an acceptable factor of safety has been exceeded. This may be due to an error in the original design or some temporary overload during construction. If the latter, then the effect of the overstress on the strength of the structure must be taken into account. The result may be a small reduction in the design factor of safety, which may be accepted when all relevant factors have been taken into account. The limit state design concepts set out in BS 8110, Parts 1 and 2 – The Structural Use of Concrete, which includes partial safety factors, can be helpful when considering such cases.

124

When a design is being checked it is important to look closely at the basic concepts and not limit the work to an arithmetical check. Details, such as depth of bearing of beams, measures taken to ensure adequate rigidity and structural continuity, particularly when precast concrete members are used, should also be carefully scrutinised. It may be prudent to consider the installation of temporary supports pending the result of a full investigation. This enables the whole problem to be reviewed in a calmer atmosphere. Reference should be made to the Report issued by the Institution of Structural Engineers, entitled 'Criteria for Structural Adequacy in Buildings', March 1976. A Report published by the Institution of Structural Engineers in July 1980, entitled 'Appraisal of existing structures', provides detailed recommendations for investigations for structural repairs.

Generally (but not always), cracks which are parallel to the main reinforcement are not as serious as cracks at right angles. However, cracks which extend right through a member should always be repaired. The use of low viscosity resins enables specialist firms to seal such cracks so that very little, if any, reduction in strength occurs. The sealing of cracks by resin injection is discussed in detail in the next section in this chapter. Cracks on their own can be repaired by a variety of methods and with a fairly wide range of materials. One of the best known techniques is crack injection using polymer resins. This is particularly suitable when the cracked member has suffered loss of strength, or when the surface chasing out of the crack is considered undesirable. However, it is unusual for structural repairs to be limited to crack injection.

5.2 CRACK INJECTION METHODS

For cracks which are deep or which pass right through the member, crack injection can provide a satisfactory solution. It is often difficult to decide whether crack injection will, in any specific case, give the better results, compared with the more orthodox technique of cutting away the concrete and repairing with a cement/sand or epoxide mortar. There are of course, many cases where crack injection is the obvious answer, but it is the borderline jobs which are difficult to decide. It is not possible to lay down hard and fast lines on which decisions can be based, but where the concrete itself is of good quality, and there has been little or no corrosion of the reinforcement, so that the only real defect is the presence of the cracks, then crack injection alone can be considered as the more suitable solution.

When rusting and spalling have occurred, then the best method of repair may be removal of the defective concrete, cleaning of reinforcement, resin injection, followed by repair with mortar. In this way, the crack within the thickness of the member is filled with resin (or at least partially filled). A final finish of the whole member with a decorative sealant would complete the job and should leave the structure in very good condition.

The essential feature of the resin injection is to inject a suitably formulated resin into the cracks. Correct formulation of the resin is of vital importance. One of the advantages of present day resins is the scope they provide for variations in formulation to obtain optimum characteristics. The requirements for the resin are likely to change from job to job.

The resins most used are epoxide, polyester and a combination of epoxide and polyurethane. The desirable qualities include low viscosity, ability to bond to damp concrete, suitability for injection in as wide a temperature range as possible, low shrinkage, and finally toughness rather than high strength (this latter means a relatively low modulus of elasticity combined with a high yield point). A low 'E' value is of particular importance when further movement is expected to take place across the cracks. Most cracks in concrete are caused by either tension or shear, and the use of a resin with a high 'E' value could result in the concrete cracking again near and probably parallel to the repaired crack.

The work of crack injection can be considered under the following headings:

Preparation of the cracks.
Location of injection points and surface sealing.
Injection of the resin.
Removal of injection nipples (if used) and plugging the hole.
Removal of the sealing strip and any final surface treatment which may be required.

5.2.1 Preparation of the Cracks

This should consist of removal of any loose weak material on the surface, followed by cleaning of the crack if this is considered necessary. Generally, cracks which are less than 0·5 mm wide are unlikely to require cleaning out unless they have been fouled by the use to which the

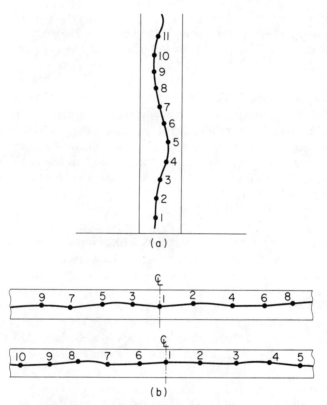

FIG. 5.1. Diagram of crack injection. (a) Injection points for vertical members. (b) Alternative orders for injection points in horizontal members.

structure has been put. Compressed air and solvents can be used for this cleaning work and for removal of water in the crack. The author favours the careful use of compressed air as this will remove dust and fine particles as well as water.

5.2.2 Location of Injection Points and Surface Sealing
The distance apart of the injection points will largely depend on the depth and width of the crack. The object is to have as few injection points as possible consistent with maximum resin penetration and ease of filling of the crack, combined with low operating pressure.

After the preparation of the cracks described above, the crack has to be sealed on its surface and the injection points marked. The sealing can be done by a variety of simple materials. The injection points can be either just holes drilled on the line of the crack or they may be nipples screwed into the concrete. Figure 5.1 shows, in diagram form, crack injection procedures for horizontal and vertical members.

5.2.3 Injection of the Resin

Crack injection with resins is a comparatively new technique and should only be entrusted to specialist firms, preferably those which formulate and use their own resins. It is natural in these circumstances that each firm develops its own technique for the job, and there is considerable divergence of views on the most appropriate method of injection and on the equipment to be used.

Some firms favour simple means of injection by gravity feed or pressure guns, the resin being premixed in batches. Others adopt more sophisticated equipment for continuous feed of freshly mixed resin and hardener through separate feed lines. One firm supplements the pressure feed by the use of a kind of vacuum mat. The author cannot really see the need of a vacuum to aid the penetration, because successful penetration has been achieved on so many occasions without it. As a matter of principle, provided the work is done properly, the simpler the equipment and less complicated the method of application, the better. In appropriate cases, the degree of penetration can be checked by coring, gamma radiography, and ultrasonic pulse velocity survey, but this checking is rather unusual.

Injection pressures are governed by the width and depth of the crack and the viscosity of the resin. The actual pressure used is low in real terms, and seldom exceeds $1\cdot0$ atm (about $0\cdot10\,\text{N/mm}^2$).

What is required with all crack injection, is uniform penetration and complete filling of the crack. A deliberately induced fluctuation in the pressure has been found to be more effective than an increase in continuous pressure. In controlling the injection process it must be remembered that the volume of resin used in filling a crack is very small.

Where the cracks are inclined or vertical, it is usual to commence injection at the lowest injection point and work upwards as the resin emerges from the next higher injection point.

For horizontal cracks, there is no fixed order of work, as the injection can start from one end and proceed along the crack, or start in the middle and work first left to the end and then right to the end, or alternatively left and right from the middle. Figure 5.1 shows in diagram form a sequence of operations for crack injection in vertical and horizontal members.

FIG. 5.2. Crack injection in large concrete manhole slab (courtesy: Cementation Research Ltd).

FIG. 5.3. Crack injection of colonnade at Victoria and Albert Museum (courtesy: Cementation Research Ltd).

5.2.4 Final Work Following Injection

It is usual to remove the injection nipples (if they are used) and seal the holes as the work proceeds. The removal of the sealing strip can be done after the resin has cured (which can be from two to seven days), or as soon as it has set which may be within a few hours.

Crack injection, particularly for fine cracks, can be useful when it is important for the repair to be as inconspicuous as possible. Even so, it will be seen. Surface grinding and some 'cosmetic' treatment will however help to mask the repair.

Figure 5.2 shows cracks in a large reinforced concrete cover into which resin is being injected as a satisfactory alternative to rejecting the unit and making a new one. Figure 5.3 shows resin injection of a colonnade at the Victoria and Albert Museum, London.

5.3 STRENGTHENING AND REPAIR

The strengthening and repair of a concrete structure or part of a structure may be required for three principal reasons:

1. When it is intended to change the use of a structure in such a way that the live (superimposed) loads will be increased to an extent that would make a fundamental change in the design necessary. Under the same heading is the case where the use has already changed and the additional loading has caused over-stress in the structural members.
2. Where the structure has for various reasons deteriorated to an extent that its members are no longer able to carry the imposed loads with an adequate factor of safety.
3. Where there is some combination of 1 and 2 above.

The first step is to inspect the structure. With isolated above-ground structures, particularly when they are in an exposed position, it may be found that there is a noticeable difference in the pattern of cracking and extent of the deterioration between the north and south sides. The south side is more exposed to rapid changes in temperature while the north may have suffered more severely from low temperatures and freeze-thaw effects.

The site inspection should include the taking of detailed notes and clear sketches together with high quality photographs which must be properly captioned and dated. The original drawings and design data should be obtained whenever possible. It is usually essential to check as far as it is practicable, the position, condition and amount of the reinforcement. The reinforcement can generally be located with a standard type electromagnetic cover meter, provided the depth to the steel does not exceed about 90 mm.

In early reinforced concrete structures unusual aggregates were sometimes used. In one case it was found that the aggregate contained ferrous particles which showed on the cover meter as steel reinforcement. Cores taken later showed that in fact there was no reinforcement at all in the slab which was supported on concrete encased steel beams.

After obtaining information about the existing reinforcement (if any), an assessment has to be made of its structural value. In some cases, due to corrosion it is prudent to ignore it and to provide new reinforcement to take all the design stresses as would be appropriate for a new structure.

Figure 5.4 shows the strengthening of a heavily loaded beam by means of reinforced gunite.

The ideal material for the repair and strengthening of a wide range of structures is high strength reinforced gunite. Gunite is a pneumatically applied material consisting of cement, aggregate and water. There are two guniting processes in general use, but of these only one, the dry-mix process, is used for most structural work in the UK. A small amount of wet-mix is used for non-structural repairs and the nozzle velocity is generally lower than that in the dry-mix method.

The cement and sand is batched and mixed in the usual way and conveyed through a hose pipe by compressed air. A separate line brings water under pressure and the water and cement/aggregate mix is passed through and intimately mixed in a special manifold and then projected at high velocity onto the member being repaired. The force of impact compacts the material, which in good quality work has a density of 2050–2150 kg/m³ (0·85 to 0·90 that of high grade concrete).

Gunite is used more extensively in the USA, Australia and South Africa than in the UK, although its use here has gained ground in recent years, both for first class repair work and as a structural material in its own right. In the USA gunite is known as 'shotcrete' and there is a Code of Practice for it, namely, 'Recommended Practice for Shotcreting', American Concrete Institute, No. ACI.506-66. In the UK there is no National Code of Practice for gunite nor for shotcrete. There is a Data Sheet on Sprayed Concrete issued by the Concrete Society and a Technical Report (No. 15) on the repair of fire damaged structures by

FIG. 5.4. View of strengthening of heavily loaded beam prior to guniting (courtesy: Cement Gun Co. Ltd).

gunite. The Association of Gunite Contractors in the UK have issued a Code of Practice for 'spraying concrete by the dry process otherwise known as gunite or shotcrete'. For repair work and structural strengthening repair by guniting is accepted by consulting engineers, local authorities and government departments.

The author feels it is unfortunate, to say the least, that gunite is not specifically mentioned as an acceptable method (a 'deemed to satisfy' method) for the protection against fire, of reinforcement and structural steelwork in the National Building Regulations, nor in BS 8110, Part 2.

The cement used in gunite can be either Portland or high alumina; some information on the latter cement is given in Chapter 1. The advantages of correctly made HAC mortar and concrete should not be overlooked where high early (24 h) strength, resistance to dilute acids, and high temperature, are important. It should be remembered that the w/c ratio of gunite is likely to be about 0·35 and this is of great importance when using HAC.

Before the gunite is applied, the old concrete must be properly prepared, cracks correctly treated and the new reinforcement fixed in position. The preparation of the old concrete consists in the removal of all defective and contaminated concrete. This can be done by percussion tools, grit blasting (wet or dry), or high velocity water jetting. The first two methods are well known, and some detailed information on the use of high velocity water jets is given in Chapter 4.

Reinforcement which is exposed by this cutting away process should be removed if it is severely corroded, or it should be cleaned by wire brushes, needle guns, grit blasting or high velocity water jets.

Cracks wider than about 0·5 mm should be cut out and filled in with either a hand applied cement mortar or with gunite. After this preparatory work has been completed, the new reinforcement is fixed in position. The method of fixing will depend on the type of structure being repaired and the detailed design of the new work.

Some examples of structures which have been repaired and strengthened with high strength reinforced gunite are given in the following sections.

5.4 THE REPAIR AND STRENGTHENING OF SILOS, BINS AND HOPPERS HOLDING FINE GRAINED MATERIALS

Silos, bins and bunkers for the storage of granular materials and constructed in reinforced concrete have been in use for more than 40 years. However, precise information on the stresses developed in the walls of these structures during filling and emptying is still lacking.

The American Concrete Institute have produced recommendations for the design and construction of silos, bins and hoppers for holding granular material. The recommendations were prepared by the ACI Committee No. 313 and published in the Journal of the ACI in October 1975. This paper includes a statement that the stresses induced in silo walls by fine grained materials during withdrawal are likely to exceed by many times the stresses caused by the same materials at rest.

A detailed account of an investigation into the cracking in a reinforced concrete cement silo is given in Technical Report No. 296, dated April 1958, by R. E. Rowe of the Cement and Concrete Association, London.

The Report concludes that the cause of the cracking was a combination of the effects of the pressure of the fine grained material and the steep temperature gradient through the walls of the silo. The stresses thus developed had not been adequately allowed for in the original design.

An example of a typical repair job follows. A group of reinforced concrete cement silos built 35 years ago were found to be in a seriously deteriorated condition. Radial cracking had developed at approximately 2·0 m centres and in some places the cracks were up to 40 mm wide. The reinforcement was badly corroded causing considerable spalling of the concrete. The cause of the trouble was diagnosed as errors in the original assessment of the working stresses, poor quality concrete and inadequate cover to the reinforcement. The silos were 17 m high, 5 m diameter, and the walls were 300 mm thick with a number of rectangular openings near ground level. The surface of the concrete was prepared by grit blasting and all cracks wider than about 0·5 mm were cut out and filled in with gunite. After the completion of this preparatory work the new reinforcement was fixed in position on the outside of the walls. This consisted of a fabric (No. A.193—BS 4483) for the complete circumference and full height of the walls; in addition, there were high tensile deformed bars, 16 mm diameter at 200 mm centres for the bottom third of the height, then 12 mm diameter at 200 mm centres for the middle third, and finally, 8 mm diameter at 150 mm centres for the top third. The rectangular openings in the base of the walls required special detailing to ensure that there was no reduction in the strength of the walls at these positions. Each opening was therefore provided with a structural steel frame consisting of two uprights and a head welded together. To these uprights high tensile deformed bars, 16 mm diameter at 200 mm centres, were welded as starter bars; these bars were 500 mm long.

The total thickness of the gunite was 100 mm and the minimum cover to the reinforcement was 25 mm. The total cost of the work was not more than 10% of the estimated cost of demolition and rebuilding.

5.5 REPAIRS TO CONCRETE DAMAGED BY FIRE

Usually after a serious fire, the appearance of the building is such that the owner is very depressed and feels that demolition and rebuilding, with all that this entails, is the only course which can be followed.

5.5.1 General
The total cost of this can be very high indeed, and a major portion of this

is the financial loss due to cessation of business from the time of the fire until the premises can be re-opened; this can be anything from six months to two years or more. It is therefore important that a realistic assessment of the damage should be made as soon as possible after the fire. When the debris, excess water, etc., have been cleared up, a careful examination of the structure will often show that despite the depressing appearance of the building, due to charred finishings, half-burnt stock and smoke blackened structure, most of the building damage is repairable and very little, if any, demolition is needed.

Concrete structures certainly suffer damage when there is a severe fire but this is appreciably less than structures of other building materials. It is sometimes stated that there are many uncertainties about the behaviour of concrete subject to fire. This is true as far as it goes, because the effect of fire on concrete depends on the temperature reached and the length of time that temperature is maintained, as well as the characteristics of the concrete in terms of cement type, w/c ratio, cement content, aggregate type, and the thickness of the concrete cover to the steel reinforcement. In the case of lightly reinforced concrete panels, the thickness of the panel and the temperature gradient through it are also important.

In recent years there have been some ill-informed statements that when polyvinylchloride (PVC) is burnt, chlorine is liberated in such quantities that when it is dissolved in the water from the sprinklers and/or fire hoses, hydrochloric acid is formed in sufficient concentration to attack and seriously weaken any Portland cement concrete with which it may come in contact. Under normal site conditions, a very weak solution of hydrochloric and hypochlorous acids may be formed, but it is most unlikely that the concentration will be at a level which would result in anything more than a slight etching of the surface of the concrete. In the majority of cases even this would not occur. On the other hand, fumes from burning PVC and other similar materials can be dangerous to health and a serious hazard to those in proximity to the fire.

There may be attack on exposed ferrous metals, but iron rusts in a humid atmosphere in any case unless it is protected. The normal concrete cover to reinforcement would give adequate protection against this form of attack.

5.5.2 Assessment of Damage
In this section the effect of fire on a building structure is being considered and such fires are of comparatively short duration, generally a matter of hours. One of the most important facts to be established as accurately as

possible, is the maximum temperature reached during the fire in various parts of the building. This applies to the exposed surface of the structural elements, namely, floors, beams and columns. In the case of floor slabs and beams it is the soffit which is normally the worst affected.

A considerable amount of testing of materials and structural elements is carried out by the Fire Research Station at Boreham Wood, Hertfordshire. The results of these tests are used for the fire grading of materials and components; these gradings are included in the Building Regulations, and expressed in terms of hours of fire resistance. There is considerable argument about the true relevance of the fire test results and the actual effect of a fire in a building on similar materials and components. There are bound to be differences between the artificial conditions surrounding a laboratory test and the conditions in a building which catches fire. Not only is the temperature different, but the concentration of flame, duration, effect of the presence of other materials, and operation of fire extinguishing equipment, all influence the effect of the fire on the building materials and the structure.

In addition to the change in strength of the materials due to the temperature reached, there will also be the stresses set up in the structural elements by the rise in temperature to the maximum reached during the fire and the return to ambient when the fire is extinguished. The thermal expansion and contraction stresses will depend on a number of factors including the restraint imposed on the elements by other members in the structure and the thermal gradient through the members. These latter are what may be termed 'site effects', and cannot be effectively simulated in laboratory tests.

However, as with most control testing, there is really no practical alternative, and laboratory testing does provide very useful information. Its limitations must however be realised. Experience has shown that in general, materials and structural elements which give good results in fire tests behave well in actual fires, and vice versa.

It is important to ascertain, as accurately as possible, the maximum temperatures reached in different parts of a structure during a fire and the temperature gradient through the structural members.

Maximum temperatures can be estimated reasonably accurately by a careful examination of the debris after the fire. It is therefore obvious that the sooner this examination is carried out the better, otherwise it may be found that cleaning-up operations have removed or destroyed valuable information. The temperatures determined in this way are likely to be appreciably different to the temperatures reached by many of the structural elements in the building.

The temperatures reached by the structural members can usually be estimated within certain limits, from a study of the visible changes which occur in concrete, and this information can then be used to decide what additional investigations and testing are required in order to determine the repairs needed.

It is pointed out in the following section that these visual effects may be misleading and should not be considered as conclusive evidence on their own. For this reason, an examination of the debris of other materials will be useful in building up a realistic assessment of the effect of the fire on the structure.

The 'other materials' mentioned above would normally be metal articles, ceramics (clayware), and glass. The time for which the materials in question have been exposed to the temperature is of course important and therefore information on the duration of the fire and its apparent intensity is most relevant.

Some examples are given below, but more detailed information can be obtained from National Building Studies, Technical Paper No. 4, Investigations on Building Fires, 1950, issued by the Building Research Station.

Aluminium and its alloys will form drops at about 600°C–700°C.
Cast iron will form drops at about 1100°C–1200°C.
Brass will form drops at about 900°C–1000°C.
Glass softens at about 700°C and melts (flows) at around 850°C.

The fundamental question is what the effect of fire is on a reinforced concrete structure. In providing an answer to this question one has to consider the effect on the constituents of concrete as well as on the steel reinforcement, and on the structure as a whole.

5.5.3 The Effect of High Temperature on Portland Cement

Clearly the effect of high temperature will depend on the temperature reached and the length of time that temperature was maintained. Consideration will first be given to the Portland cement paste, because if this is seriously weakened the concrete is likely to disintegrate.

The uncombined (surplus) water in the concrete will be driven off from the surface layers and some shrinkage cracking in those layers will occur. Up to a temperature of about 100°C there will be no significant loss of chemically combined water even with a prolonged exposure to this temperature. As the temperature rises above 100°C there is a gradual loss of the chemically combined water from the calcium silicate hydrates. The actual loss depends on time and temperature. With this loss of chemically combined water there is a drop in the strength of the concrete corresponding to the amount of water lost. However, once the concrete cools down there will be no further reduction in strength.

If the temperature of the concrete reaches 400°C and above, the calcium silicates commence to decompose into quicklime and silica. This process is irreversible and there is a progressive loss of strength with time. When the concrete cools, the quicklime (calcium oxide) will absorb moisture, converting to slaked lime (calcium hydroxide). When this happens, disintegration of the affected concrete will occur.

5.5.4 The Effect of High Temperature on High Alumina Cement

With temperatures above 100°C, there is a progessive loss of chemically combined water from the calcium aluminate hydrates. At about 400°C decomposition sets in but this produces mostly calcium aluminate and alumina which are much more stable than the calcium oxide produced by decomposition of Portland cement. This is basically why HAC is a refractory cement, and, with special aggregates, makes refractory concrete.

5.5.5 The Effect of High Temperature on Aggregates

The effect of temperature will depend on the mineralogical classification of the aggregate. For the purpose of this discussion, the aggregates used for structural concrete can be divided into natural and artificial aggregates. The former group is subdivided into igneous, siliceous, and calcareous aggregates.

Igneous rocks, such as granite, basalt, etc., are likely to be reasonably stable up to about 1000°C, although there may be some spalling with a rapid rise and fall in temperature.

Siliceous aggregates, such as flint gravels, undergo rapid changes in volume at certain temperatures and this sudden increase in volume may cause disruption of the concrete. The critical temperatures are about 250°C and 575°C.

Calcareous aggregates undergo fairly uniform expansion with increase in temperature. At about 400°C they tend to change to form quicklime (as is the case with Portland cement). On cooling. calcium hydroxide is formed with the absorption of moisture with a serious reduction in strength.

Artificial lightweight aggregates, such as expanded clay and shale, sintered pulverised fuel ash, and blastfurnace slag, are manufactured at temperatures above 1000°C and are therefore very stable at temperatures below this level.

5.5.6 The Effect of High Temperature on Steel Reinforcement

The effect of temperatures up to about 700°C on the final strength and ductility of mild steel and hot-rolled high yield steel is from a practical point of view negligible. This refers to the strength and ductility after return to ambient temperature. The effect of the elevated temperature on the reinforcement under load and the disruptive effects of expansion must be given careful consideration.

5.6 INVESTIGATIONS AND TESTS PRIOR TO REPAIR

In the following paragraphs recommendations are made for the overall investigation of a concrete building structure which has been damaged by fire. The taking and testing of cores and samples of steel reinforcement is suggested in order that a realistic appraisal be made of the condition of the structure and clear recommendations made for repair. The sampling and testing should be done with discretion as it is very expensive, and a great deal of very useful information can be obtained by careful visual examination alone.

After consideration of the likely effect of high temperatures on Portland cement, aggregates and steel reinforcement, this information must be given practical effect by its application to reinforced concrete.

The exposed surface will obviously receive the full heat of the fire and the temperature will reduce sharply as the distance from the face increases. With temperatures above 400°C permanent damage is done to the cement paste and therefore this can be considered as applying to the concrete as well. The concrete cover to reinforcement is reasonably effective as thermal insulation and it has been found that in what may be termed 'normal' fires, the temperature of the concrete at a depth of 50 mm from the exposed face, usually does not exceed about 300°C. In very severe fires, this temperature may be reached at a depth of 75–100 mm. From what has been said, it can be seen that the critical temperature for the concrete is about 400°C, when the calcium silicates in the cement start to be converted into calcium oxide. Many experienced engineers take this critical temperature rather lower, namely at 300°C.

It is perhaps fortunate that at this 'critical' temperature, whether it be taken as 300°C or 400°C, there is a noticeable colour change, consisting of a pink or pale red coloration. Provided there is adequate lighting, which should in fact always be provided for examination purposes, this pink colour can be easily seen. However, the pink colour may disappear if the temperature reached was very high, but in such a case, the concrete would be very friable. It has also been reported that this colour change sometimes disappears with the passage of time. It is therefore important for investigations to start as soon after the fire as possible.

It should not be assumed that final decisions on the extent of the damage can be taken from visual inspections alone. The visual examination must be supplemented by cutting back the concrete with hand or pneumatic tools. This quickly discloses concrete which has suffered a reduction in strength. In addition, some cores should be taken, and the newly developed ultrasonic pulse velocity (upv) technique has proved very useful.

With large fires, it is always found that some of the reinforcement in the slabs has buckled and this means it must be replaced. Samples can be taken from the buckled bars and tested for yield point and ductility. It is usually found that damage to the reinforcement in slabs is much greater than in beams and this is no doubt due to the greater depth of cover in beams compared with slabs.

It is essential that a carefully thought out system of recording and reporting on the damage to all parts of the structure should be adopted. This should classify the damage into a number of categories such as insignificant, medium and severe. Each category must be clearly defined by reference to basic criteria such as surface appearance (spalling, colour, cracking, crazing, etc.), estimated temperature reached, physical characteristics revealed by use of hammer and chisel and condition of reinforcement. To these can be added, as they become available, the results of core tests, tests on the reinforcement, and if appropriate, upv survey. Each serious fire produces its own set of problems and many papers have been written on the effect of fires of varying intensity in a wide range of structures. Some of these papers are listed in the Bibliography at the end of this chapter.

5.7 THE USE OF GUNITE FOR FIRE DAMAGE REPAIRS

While the normal method of repair is with reinforced gunite, each job brings its own on-site problems and difficulties. A fundamental question which arises immediately is whether a concrete structure damaged by fire and repaired with gunite (a pneumatically applied cement/sand mortar known as 'shotcrete' in the USA and 'Torcrete' in Germany) will have the same structural factor of safety and fire resistance after the repair as it had before.

In the opinion of the author, the answer, in general terms, is in the affirmative, provided the repairs are properly carried out, and the gunite is of the highest quality. It is normal practice for loading tests to be carried out to 'prove the design' and check the quality of the repair. For these tests the test load is usually 25% above the design live load. Regarding the fire resistance of cement/sand mortar, the following table taken from Code of Practice CP 114—The Structural Use of Reinforced Concrete in Buildings, shows that cement/sand mortar has at least the same fire resistance in hours as an equal thickness of concrete. Some authorities consider that high quality cement mortar resists fire better than concrete as it is less liable to spalling.

TABLE 5.1
Fire resistance of reinforced concrete columns
(from Code of Practice CP 114, Table 29 B, p. 90)

Additional protection	Minimum dimensions of concrete in mm for a period of					
	4 h	3 h	2 h	$1\frac{1}{2}$ h	1 h	$\frac{1}{2}$ h
None	450	400	300	250	200	150
Cement or gypsum plaster 12 mm thick on light mesh reinforcement fixed around the column	425	375	275	225	175	150

This table clearly indicates that 12 mm of cement plaster applied on a mesh which is fixed to the column, has the same fire resistance as 12 mm of original thickness of the concrete in the column.

The recommendations in BS 8110—The Structural Use of Concrete, Part 2 (BS 8110, published in 1985 superseded CP 110:1972) are different to those in CP 114. Codes of Practice have no legal authority on their

FIG. 5.5. Underside of hollow pot and concrete beam floor damaged by fire
(courtesy: Cement Gun Co. Ltd).

own and are guides for good practice. The legal requirements for fire
resistance are contained in the Building Regulations 1985 which replaced the Building Regulations 1976.

For all readers who may be responsible for the repair of fire damaged
concrete structures it is essential to read carefully the Building
Regulations 1985, the Building Research publication, Guidelines for the
construction of fire resisting structural elements (HMSO, 1982) and
BS 8110, Part 2, Section 4. It is important to note that the Building
Regulations refer to BS 8110, and BS 8110 refers to the BRE 'Guidelines'.
The author's interpretation of clause 4.2.4 in Part 2 of BS 8110 is that the
thickness of cement/sand mortar used to provide effective cover to

FIG. 5.6. View of underside of floor in Fig. 5.5 during repair with gunite (courtesy: Cement Gun Co. Ltd).

reinforcement can be 0·6 times the thickness required for concrete cover, provided the thickness does not exceed 25 mm.

The use of gunite for the repair of fire damaged reinforced concrete structures is well known the world over. Figures 5.5–5.8 show typical fire damage and how this has been successfully repaired by the use of gunite.

5.8 REPAIRS TO CONCRETE BEAMS USING BONDED STEEL PLATES

The possible use of bonded steel plates to the soffit of reinforced concrete beams in order to strengthen them was developed in the early 1970s at

Fig. 5.7. Soffit of warehouse floor badly damaged by fire (courtesy: Cement Gun Co. Ltd).

FIG. 5.8. View of floor in Fig. 5.7 after fixing of new rebars prior to guniting (courtesy: Cement Gun Co. Ltd).

the Transport and Road Research Laboratory, Crowthorne. The intention was to use this to increase the load carrying capacity of highway bridges to meet increased traffic loadings or where the bridge, for some reason or other, had inadequate strength. This technique is mentioned in rather more detail in Chapter 8, on Bridge Repairs. The intention here is to draw the attention of readers to its limited use in strengthening floors in building structures.

It appears that probably the first use of this technique was in France and S. Africa. Since then it has been adopted successfully in Switzerland and Japan.

Work at TRRL, which is documented in the Bibliography at the end of this Chapter, indicates that certain factors are important, and these may be summarized as follows:

(a) Great care must be taken in the preparation of the concrete to receive the resin adhesive, and the same applies to the steel plates.
(b) The adhesive is likely to be an epoxide resin and this should be carefully selected.
(c) The whole of the steel plate surface in contact with the concrete must be covered with the adhesive.
(d) The strength of the cured resin is likely to be higher than that of the concrete to which it is bonded so that if failure occurs it will occur in the concrete. The tests at TRRL confirmed this.

A basic fact which must be given proper consideration is that epoxide resins are much more sensitive to fire than concrete, and therefore the technique is only suitable for structures where the fire risk is small. Protection by gunite would increase the dead load of the beam.

The steel plates and the concrete beam must act compositely which requires that the bond between them must be as strong as it is practical to make it.

BIBLIOGRAPHY

AMERICAN CONCRETE INSTITUTE, Commentary on recommended practice for design and construction of concrete bins, silos, and bunkers for storing granular material, Committee 313, Paper 72-38, *Journal A.C.I.*, October 1975, pp. 549–565.

AMERICAN CONCRETE INSTITUTE, *Concrete repair and restoration* (ACI, Compilation No. 5); ref. C-5, 1980, p. 118.

AMERICAN CONCRETE INSTITUTE, *Guide for determining the fire endurance of concrete elements*, ACI Committee 216; ref. 216R-81, 1981, p. 36.

AMERICAN CONCRETE INSTITUTE, *Application and use of shotcrete*, ACI, Compilation No. 6; ref. C-6, 1981, p. 92.

ASSOCIATION OF GUNITE CONTRACTORS, *Code of Practice for spraying of concrete by the dry process, otherwise known as gunite or shotcrete*, published by the Association, London, p. 4.

BATE, S. C. C., *Structural failures*, Paper given at Joint Building Research Establishment and Institute of Building Seminar, Nov. 1974, p. 11.

BRITISH STANDARDS INSTITUTION, *Guide to the assessment of concrete strength in existing structures*, BS 6089:1981.

BRITISH STANDARDS INSTITUTION, *The structural use of concrete*, Parts 1 and 2, 1985 (replaces CP 110).

BUILDING RESEARCH ESTABLISHMENT, *Guidelines for the construction of fire resisting structural elements*, HMSO, London, 1982.

BUILDING RESEARCH ESTABLISHMENT, *Fire performance of walls and linings*, Digest 230, HMSO, London, 1984.

CATON CROZIER, A., Strengthening on fifty year old viaduct, *Concrete*, July 1974. Reprint, p. 5.

CIRIA, *Spalling of concrete in fires*, Technical Note 118, published by CIRIA, London, 1985.

CONCRETE SOCIETY, *Concrete core testing for strength*, Technical Report No. 11, May 1976, p. 44, published by the Society, London.

CONCRETE SOCIETY, *Guide to precast concrete cladding*, Technical Report No. 14, 1977, p. 24.

CONCRETE SOCIETY, *Assessment of fire damaged concrete structures and repair by gunite*, Technical Report No. 15, 1978, p. 28.

CONCRETE SOCIETY, *Durability of tendons in prestressed concrete: Recommendations on design, construction, inspection and remedial measures*, Technical Report No. 21, 1982, p. 8.

CONCRETE SOCIETY, *Repair of concrete damaged by reinforcement corrosion*, Technical Report No. 26, 1984, p. 32.

DORE, E. The assessment of fire damage to concrete structures. *Concrete*, Sept. 1984, 49–51.

ELDRIDGE, H. J., *Diagnosing building failures*, Paper given at Joint Building Research Establishment and Institute of Building Seminar, Nov. 1974, p. 13.

FREEMAN, I. L., *Failure patterns and implications*, Paper given at Joint Building Research Establishment and Institute of Building Seminar, Nov. 1974, p. 13.

GREEN, J. K., Some aids to the assessment of fire damage, *Concrete*, Jan. 1976, 14–17.

GREEN, J. K. AND LONG, W. B., Gunite repairs to fire damaged concrete structures, *Concrete*, April 1971, Reprint, p. 6.

HEWLETT, P. C. AND WILLS, A. J., *A fundamental look at structural repair by injection using synthetic resins*, Symposium on Resins and Concrete, The Plastics Institute and Institution of Civil Engineers, Newcastle on Tyne, April 1973, Paper No. 17, p. 12.

INSTITUTION OF STRUCTURAL ENGINEERS, *Criteria for structural adequacy in buildings*, March 1976, p. 35.

INSTITUTION OF STRUCTURAL ENGINEERS AND CONCRETE SOCIETY, *Fire resistance of concrete structures*, August 1975, p. 59.

LETMAN, A. J. AND HEWLETT, P. C., Concrete cracks—a statement and remedy, *Concrete*, Jan. 1974, Reprint, p. 5.

LORMAN, W. R., *The Engineering properties of shotcrete*, American Concrete Institute, Publication SP14A, Dec. 1968, p. 58.

RAITHBY, K. D., *External strengthening of concrete bridges with bonded steel plates*, Transport and Road Research Laboratory, Supplementary Report 612, 1980, p. 18.

SOLOMAN, S. K., SMITH, D. W. AND CUSENS, A. R., Flexural tests of steel-concrete-steel sandwiches. *Mag. Conc. Res.*, **28** (94), March 1976.

Chapter 6

The Repair of Concrete Floors and Roofs

The deterioration of concrete floors and problems of leaking flat roofs are unfortunately more prevalant than defects in other parts of building structures. It is difficult to find a logical and satisfactory explanation for this, but the problem of fairly widespread defects in both floors and roofs is a fact that cannot be ignored.

6.1 THE REPAIR OF CONCRETE FLOORS

The repairs, which for this purpose include resurfacing, are dealt with under the following headings:

1. Patching and small scale repairs.
2. Repairs to joints.
3. Repairs to cracks.
4. Resurfacing with insitu bonded toppings.
5. Resurfacing with insitu unbonded toppings.
6. Resurfacing with precast concrete flags.
7. Resurfacing with ceramic tiles.
8. Resurfacing with chemically resistant tiles and bricks.
9. Resurfacing with extra heavy duty PVC.
10. Resurfacing with semi-flexible insitu toppings.

6.1.1 Patching and Small Scale Repairs
The careful patching of small defective areas can provide a satisfactory answer to a deteriorated concrete floor slab. This type of repair should generally be carried out in the following way:

(a) All weak and defective concrete should be cut away, and all grit and dust removed. The patches should be cut out as square as practicable.

(b) The surface of the concrete around the cut out areas should be cleaned and wire brushed for a distance of 50 mm.

(c) The surface of the cut out concrete should be well wetted (preferably overnight) and then, not more than 20 min before the new concrete or mortar is laid, a coat of Portland cement/SBR grout should be well brushed into the prepared surface of the cut out concrete.

(d) The concrete/mortar must be well compacted and finished with a wood or steel trowel to give the required texture.

(e) After completion of trowelling, the repaired areas must be either sprayed with a resin-based curing membrane, or covered with polyethylene sheets held down around the edges and kept in position for 4 days.

The mix proportions for the concrete and mortar are likely to be about:

Concrete: 1 part ordinary or rapid hardening Portland cement.
$2\frac{1}{2}$ parts concreting sand.
$2\frac{1}{2}$ parts 10 mm coarse aggregate.
The amount of mixing water should be such as to give a slump of 50 mm $\pm$ 25 mm which should be adequate for the compaction of small areas of concrete.

Mortar: 1 part cement (as above).
3 parts coarse concreting sand.

The proportions given above are for volume batching as it is assumed that only small amounts of concrete/mortar will be used.

It is recommended that a styrene–butadiene (SBR) latex emulsion be used as part of the gauging liquid (to replace some of the water). The usual proportions for the SBR are 10 litres of emulsion to one 50 kg bag of cement. The use of the SBR will improve bond to the old concrete, reduce shrinkage and reduce permeability and improve chemical resistance.

When ordinary Portland cement is used, all traffic should be kept off the repaired areas for at least 7 days; this can be reduced to 2 days if RH Portland cement is used. In cold weather these periods should be extended by about 50%. Heavy traffic should not be allowed over the new concrete for 14 days with OPC and 10 days for RHPC.

There are occasions when emergency repairs have to be carried out over a weekend, so that the floor can be put back to full operation on the Monday morning. In such cases, high alumina cement can be used to advantage (see Chapter 1). The mix proportions recommended above should then be changed to 1 part cement, 2 parts concreting sand and 2 parts of 10 mm coarse aggregate. With the use of HAC it is essential that the water/cement ratio should be kept below 0·40, that is not more than 20 litres of water per 50 kg bag of cement when using dry sand. The addition of the SBR is particularly important as this acts as a plasticiser and so allows a reduction in the amount of mixing water.

Reference has been made above to the use of concrete or mortar; it is better to use concrete where the depth of the patch exceeds about 40 mm.

As long as possible after the completion of the repair, a coat of cement/SBR slurry should be well brushed into the surface of the repaired area and the 50 mm wide strip of cleaned existing concrete around the perimeter of each patch.

If HAC is used for the repair, then it must also be used for the SBR slurry. Portland cement should not be mixed with high alumina cement (the exception being certain types of temporary work).

6.1.2 Repairs to Joints

Floors which carry moving loads from trolleys and fork-lift trucks usually show deterioration at joints before wear appears on the surface within the bays. This deterioration takes the form of fretting and ravelling along the edges of bays and along the wider cracks.

Careful investigation is required before repairs are put in hand as there are many factors involved which will influence the method of repair. These factors include:

The use to which the floor is put, particularly whether the trade is 'wet' or 'dry'.
The weight of any moving loads and the type of wheels, i.e. nylon, steel, pneumatic etc.
The temperature range within the building.

The author's experience is that many joints are not required at all once the initial contraction and shrinkage of the concrete has taken place. The provision of a comparatively wide sealing groove is often unnecessary. The groove can be reduced to 6–10 mm wide, or in some cases abolished altogether by filling with an epoxide mortar trowelled off smooth, level with the surface of the adjoining bays.

FIG. 6.1. Deteriorated movement joint in a car park floor.

Subject to a decision on whether to 'lock' the joint as mentioned above, or repair it so that movement can take place, the following method of repair will generally prove satisfactory:

(1) The ravelled, broken and otherwise defective concrete along both

sides of the joint should be cut away with a concrete saw.

(2) All sealing compound and back-up material in the joint groove should be removed and the joint carefully cleaned out.

(3) The sides of the joint should be remade with an epoxide mortar, leaving a gap of predetermined width (unless it has been decided to lock the joint as described above). Full movement joints are usually about 20 mm wide, while the groove for a contraction joint need not exceed 3 mm in width.

(4) After the epoxide mortar laid in (3) above has hardened the full movement joint should be sealed with a preformed Neoprene or EPDM cellular strip which is inserted into the joint with a primer-adhesive. Other materials which can be used are two-part polysulphides, two-part polyurethanes and epoxy-polysulphides.

It is important that the new back-up material inserted in the joint adequately supports the sealant. A suitable material is resin bonded cork. For the narrow contraction joints, thin strips of resin bonded cork can be used.

It must be remembered that all joints are liable to cause trouble due to 'fretting' at the edges, and to loss of sealant. Figure 6.1 shows a deteriorated joint in a car park floor. There are differences of opinion regarding the width/depth ratio of joint sealants. This ratio depends on the type of sealant selected and generally those given in Table 6.1 are considered reasonable.

TABLE 6.1

Type of sealant	Width:depth ratio
Bituminous and rubber/bitumen	1:1
2 part polysulphide	1:1 to 2:1
2 part polyurethane	2:1
Silicone rubber	2:1

The above is intended as a guide, and users would be prudent to follow the recommendations of the manufacturer of the sealant they propose to use.

There may be conflicting requirements for the sealant; for example, it may be desirable for the sealant to be able to accommodate appreciable movement, be resistant to certain chemicals and at the same time give good support to the sides of the joint to help prevent fretting as heavy moving loads pass across the floor.

6.1.3 Repairs to Cracks

The presence of cracks in a concrete floor can be misleading as to their effect on the durability of the floor, that is, its ability to perform the function for which it was intended. It is therefore necessary to investigate the cracking to determine the approximate width and direction of the cracks and their cause, and then to decide what should be done about it, if anything.

The presence of a limited number of fine cracks in a concrete ground floor slab used for a 'dry' trade, may well not need any repair. They are likely to cause less damage in the long term than joints, as the sides of the cracks are less liable to travel. On the other hand, cracking in a suspended floor slab may signify some structural inadequacy. Again, serious cracking in a floor of a building which will be used for the production of foodstuffs may mean that a completely new impervious topping is needed. The provision of toppings is dealt with in the next section.

The main reasons for the repair of cracks may therefore be set out as follows:

(a) To prevent fretting and ravelling along the sides of the crack (this is only likely to occur in what may be termed 'wide' cracks, i.e. wider than about 1 mm).
(b) To prevent the penetration of water or other liquids.
(c) To seal completely and securely the cracks prior to the laying of a thin bonded topping on the floor. This can only be done success-fully when the cracks are not 'alive', that is when further movement across the crack is not anticipated.
(d) Cracking in a thin bonded topping which is accompanied by loss of bond adjacent to the cracks will need careful assessment, and generally repairs will be required.

In determining the method of repair it is necessary to decide whether the crack is 'alive', that is, whether it will continue to move after the repair has been completed. If the answer is in the affirmative then the material used to repair the crack must be sufficiently flexible to accom-modate this movement; at the same time the sealant must bond to the concrete. If no further movement is expected then the repair material can be rigid.

Flexible materials include rubber bitumens, specially formulated epo-xide resins and polyurethanes, and natural and artificial rubber latexes. Care must be taken in the selection to ensure, as far as this is possible,

FIG. 6.2. Crack in concrete floor being cut out prior to sealing (courtesy: Errut Ltd).

that the material will be durable under the operating conditions of the floor. When the cracks are very narrow, less than about 0·25 mm, the crack filler should be of low viscosity.

For cracks up to about 1·0 mm wide, there is generally no need to cut out the crack. The recommended procedure is to tap lightly along the crack with a chisel, clean out all grit and dust with compressed air, and then brush into the crack a cement latex grout, or a polymer resin.

For wider cracks, particularly when the edges have spalled, the crack should be cut out. After cutting out, the crack should be carefully cleaned as described above. If a cement/sand SBR latex mortar is used, it is advisable to wet the crack overnight, and to cure the mortar for four days. About a month later it may be found that a fine hair crack has developed along each side of the repair. This can then be simply grouted-in with a cement/latex grout applied by brush in two coats.

When it is important that the floor is watertight, as may be the case with 'wet' trades, reference should be made to Chapter 9, the section on repairs to floors of water retaining structures.

Figure 6.2 shows a crack being cut out using special equipment.

6.1.4 Bonded Toppings

Cases often arise where it is necessary to provide a completely new wearing surface to the whole or major part of an existing floor. Such a major repair would take the form of either a thin bonded topping, or a thick unbonded topping. The use of the former is more usual than the latter due mainly to the need to conform as far as possible with the levels of adjoining floor areas. The use of relatively thick unbonded toppings is dealt with in Section 6.1.5. Generally a thin bonded topping can be used when the base concrete is high quality, is not contaminated with oil or grease or chemical compounds likely to be aggressive to the new topping or to interfere with the bond at the interface, and is relatively free from cracks. It must be emphasised that thin bonded toppings rely, for satisfactory performance, on very good bond with the base concrete, a good quality concrete base slab, and on the quality of the topping concrete itself.

For the reasons given below, the bay sizes of the bonded topping should conform with those of the base concrete. There are differences of opinion about the maximum thickness which should be accepted for bonded toppings. There is no doubt that cases have been reported of failures of bonded toppings which exceed about 50 mm in thickness. The reason given for this is that there is a steep moisture gradient through the new slab resulting in appreciably greater shrinkage in the top than in the bottom and this causes loss of bond and curling along the sides of the bays.

However, there are occasions when it is necessary to provide high quality abrasion resistant toppings subjected to heavy moving loads and the thickness is limited, for reasons of levels, to say 75 mm which would be too thin for an unbonded topping. The solution is undoubtedly difficult, but the author feels that the solution to the problem lies in drastically reducing the water content of the mix, say to a water/cement ratio of 0·30–0·35 by the use of a superplasticiser, to give a slump of about 50 mm with a cement content of 400 kg/m^3.

The dangers inherent in the use of these rather thick bonded toppings must be recognised and a prudent engineer would inform his client accordingly. The chances of loss of bond and curling due to drying shrinkage can be reduced by careful mix design and good control on site, but cannot be eliminated.

There should be no special problems with bonded toppings having thickness in the range 25–50 mm, but great care, quality control and attention to detail, are all necessary. For normal bonded toppings, a minimum thickness of about 25 mm is needed, but with polymer modified toppings (using styrene–butadiene or acrylic resin latex), the usual thickness (used by specialist firms) is about 12–15 mm.

Apart from the special requirement for chemical resistance to compounds which are aggressive to Portland cement concrete, the principal requirement for thin bonded toppings is abrasion resistance and the formation of a minimum amount of dust under what may be heavy traffic from fork-lift trucks with hard nylon or polyurethane tyres.

Many proprietary floor toppings are described in technical sales literature as 'jointless and dustless'. In the opinion of the author, all floor toppings of any significant area have joints as they have to be laid in bays. Also, toppings based on Portland cement, even though they are modified with a polymer, are not dustless, although the amount of dust may be small and unimportant except in special circumstances. A further point which has to be kept in mind is that where the topping passes over joints in the concrete base slab, reflected cracking may occur. This is certain to happen if, under heavy moving loads, one bay moves up or down relative to an adjacent bay; such movement need only be very small to cause the topping to crack more or less in line with the joint below.

The following are the essential requirements for a successful thin bonded topping:

(a) Preparation of the base concrete. If at all possible, the concrete should be scabbled with a mechanical scabbler. Some scabblers are multi-headed and are percussion tools; others, use hard steel shot of various diameters with a vacuum attachment to retain the dust. Both can be used successfully on concrete with a hard aggregate, such as granite or flint gravel. The author prefers the multi-headed scabbler.

All loose grit and dust must be completely removed, and this may require the use of an industrial vacuum cleaner. A thin layer of dust will have a serious effect on the bond between the base concrete and the new topping and can result in a major failure of the topping.

(b) Mix proportions. For work of importance, batching must be by weight. Cement: ordinary or rapid hardening Portland; minimum cement content, 400 kg/m³. In some cases this may have to be increased to 450 kg, see below under characteristic strength. Aggregates: a natural sand, graded medium or coarse, and to BS 882:1983. Coarse aggregate (granite or flint gravel) to BS 882: 1983; the size will depend on the thickness of the topping. The author favours 10 mm for toppings up to a thickness of 50 mm. The proportion of fine to coarse aggregate is best determined by trial mixes.

The free water/cement ratio must be strictly controlled and should not be permitted to exceed 0·40. Adequate workability to ensure proper compaction with the equipment being used is essential if a dense, strong, well bonded topping is to be provided. A low w/c ratio of 0·40 or less is almost certain to require the use of a plasticiser. A slump of 75 mm ± 25 mm should be adequate but much higher slumps can be used with advantage as this will ease compaction and placing, but the w/c ratio of 0·40 must not be exceeded.

(c) The characteristic strength of 40 N/mm² is recommended for good general work, and this should be achieved without difficulty using the mix proportions outlined above. A strength of 50 N/mm² may be needed for floors with light abrasion resistance, and this may require a cement content of 450 kg/m³ or more. With the quality of concrete recommended here, trial mixes are necessary, with strict site control of materials and workmanship.

(d) After compaction, the finishing of the topping can be carried out by power float or hand trowelling.

(e) The topping must be cured for at least 4 days by means of a good quality curing membrane or by covering with polyethylene sheets held down around the perimeter by boards or blocks.

While abrasion resistance is closely related to compressive strength, the method of finishing and quality of the curing has an important effect on the wearing properties of the topping. There is no recognised on-site method of test for abrasion resistance, but with experience, the Schmidt Rebound Hammer can be used to obtain a good indication of the likely abrasion resistance of an insitu concrete slab. Reference can usefully be made to the paper by Ralph Chaplin, and other publications listed in the Bibliography at the end of this chapter.

FIG. 6.3. View of well scabbled concrete floor (courtesy: Errut Ltd).

In addition to a high cement content (400 kg plus) and a low water/cement ratio (0·40 and lower), other methods can be used to increase abrasion resistance. These include finishing the concrete while it is still plastic with a sprinkle of finely divided iron or very hard grit which is usually premixed with cement. The weight of iron or other selected metal is generally in the range 4–12 kg/m², according to the use to which the floor will be put. The application of surface sealants and low viscosity resins in the form of very thin coats which are less than 1 mm thick, while very useful to reduce dusting, does not contribute much to wear resistance, except that due to foot traffic.

The addition of polymers such as SBRs and acrylics appears to increase wear resistance, and they certainly allow toppings to be successfully laid thinner than the 25 mm which is the usual limit for plain cement mixes.

FIG. 6.4. Warehouse floor finished with polymer modified thin bonded concrete topping (courtesy: Residura Ltd).

However, great care has to be taken with the bond to the base concrete and to all aspects of the work. The author considers that 12 mm is the absolute minimum thickness for such toppings, and 15 mm is better.

The usual weight of SBR in the topping mix is 15–20% by weight of cement (7·5–10 litres of latex emulsion to a 50 kg bag of cement).

Figure 6.3 shows the surface of a well scabbled concrete floor. Figure 6.4 shows a warehouse floor finished with a polymer modified thin bonded concrete topping.

The author feels that a word of warning is needed if it is proposed to remove an existing defective bonded topping and to replace it with a new thin bonded topping. It can prove very difficult to remove such old toppings even though they give the impression of being badly cracked with many unbonded areas. When the topping comes to be removed it is often found that while some sections come up easily, others require special equipment to remove them, which not only causes considerable delay but can cause serious damage to the existing concrete base slab on which the new topping is to be laid.

6.1.5 Unbonded Insitu Concrete Toppings

There are two main reasons for using unbonded toppings on existing concrete base slabs:

(a) Because, for various reasons, the topping will be thick enough for bond to the base slab to be unnecessary.

(b) Due to unacceptable bay sizes and layout in the base slab and/or numerous cracks which may still be 'alive', a thick topping separated from the base slab is necessary to avoid reflected cracking and/or to provide a satisfactory bay layout.

The separation between the old and new concrete can be achieved by the provision of a slip membrane.

The mix proportions, characteristic strength, compaction, finishing and curing should be as recommended for thin bonded toppings. However, because the topping is acting independently of the base slab, which in fact acts as a sub-base, the unbonded topping should be considered as a new concrete ground floor slab and designed and constructed accordingly, as far as bay layout and provision of reinforcement is concerned. If the unbonded topping is laid on a suspended slab, then special attention must be paid to the way the suspended slab has been designed. Consideration of these design conditions is outside the scope of this book, and advice from an experienced civil or structural engineer is essential. When the new unbonded topping is cast on an existing ground floor slab, the bay width is usually about 4.5 m (to fit the width of standard fabric reinforcement to BS 4483), and the lengths of the bays are governed by the weight of reinforcement in the longitudinal direction. Details for design are given in the Bibliography at the end of this chapter.

Reference has been made at the beginning of Section 6.1.4 to the problem of providing toppings to existing concrete floors in the range 50 mm to say 100 or 125 mm. The minimum thickness for an unbonded topping will depend mainly on the load which it has to carry. The author recommends a minimum of 100 mm, with a decided preference for 150 mm and upwards.

Problems can arise when the use of a floor is changed from a dry to a wet trade and the new topping has to be laid to falls to drainage channels. To achieve this, the thickness of the topping can change from say 150 mm at one side to about 50 mm at the floor channel. If the whole topping is laid as unbonded, there is a distinct chance that an irregular crack may form more or less parallel to the channel and located where the thickness of the unbonded topping is insufficient to carry the loads imposed on it. On the other hand, if the thinner part of the topping is bonded to the base slab while the thicker part is left unbonded, the change in conditions of restraint may also cause cracking. There is no entirely satisfactory solution; if it is considered important to avoid a random crack parallel to the channel, then a plain butt joint should be formed in the topping at a selected distance from the channel. If this joint opens then it can be sealed.

6.1.6 Repairs with Precast Concrete Flags (Slabs)

The use of hydraulically pressed concrete flags can provide a new and durable surface to a deteriorated concrete floor. The flags should comply with BS 368–Precast Concrete Flags, as this should ensure that they are abrasion resistant, strong, and relatively impermeable, provided they are specified as 'pressed' in accordance with clause 6.1 of the Standard. The concrete floor slab should first be cleaned and repaired so as to provide a reasonably clean, level surface on which the flags can be laid.

The mix for the bedding mortar should be rather stiff, with proportions of 1 part ordinary Portland cement to 3 parts of clean sand, grading M or F, Table 5, BS 882:1983. The thickness of the bed should be in the range 15 mm to 40 mm, and the flags must be fully bedded, that is, the whole area of the underside of the slabs must be in contact with the mortar into which the flags should be well punned. The flags must be carefully laid to ensure that there is no lipping at the joints; with moving loads on hard wheels, any unevenness at the joints will quickly lead to deterioration of the edges of the flags. The use of an SBR emulsion in the mix, 10 litres to 50 kg cement, is recommended for both the bedding mortar and the jointing mortar.

The joints should be as narrow as practicable, 3 mm, and must be completely filled with the jointing mortar. The mortar for bedding can be either ready-mixed, or mixed on site in a pan or forced-action type mixer. Tilting drum type mixers are not suitable for mixing cement/sand mortars.

With such narrow joints (3 mm) there is likely to be difficulty in ensuring that the joint gap is completely filled, and joints are the weakest part of the floor. Two methods can be used;

(a) To brush in dry cement and fine sand and then lightly water the joints as they are filled.

(b) To use a grout composed of cement and fine sand and an SBR emulsion with sufficient water to make the grout flow. This is then thoroughly worked into the joints until they are completely filled (this is the technique used for ceramic tiles).

With the first method it is difficult to fill the joints completely. With the second method, the surface of the flags will become stained with the grout.

The concrete flags are manufactured in four standard sizes and each size is produced in two thicknesses, 50 mm and 63 mm. For resurfacing a deteriorated concrete floor, the 63 mm thick flags should be used. The standard sizes are 600 mm × 450 mm; 600 mm × 600 mm; 600 mm × 750 mm and 600 mm × 900 mm. To keep the number of joints to a minimum it is generally better to use the 600 mm × 900 mm flags.

6.1.7 Resurfacing with Tiles

The tiles for heavily loaded floors can be terrazzo, high quality pressed concrete, or ceramic. It is the author's opinion that the method of laying the tiles, including the preparation of the base concrete, should be basically the same for these three types of tiles. The concrete base slab should be repaired and the sound surface scabbled or grit blasted to form a good key for the bedding mortar; the repaired areas should be left with a rough, laitance-free surface. It is important that all grit and dust should be removed and the surface damped down overnight prior to the laying of the bedding mortar.

To ensure uniformity of mix proportions of the bedding mortar, the mortar should either be ready-mixed or weigh batched on site or pre-bagged. Mixing should be in a pan or forced action type mixer, as the tilting drum type concrete mixer does not cope well with cement/sand mixes. Volume batching on site may be permitted if gauge boxes are used and there is good site control.

The sand should be a clean concreting sand complying with grading C or M (or some combination of both) of Table 5 of BS 882:1983. The mix proportions should be in the range of 1:3 to 1:3·5 by volume. The amount of water in the mix is critical for ensuring strength and workability. Thorough compaction of the bedding mortar is absolutely essential if the floor is to stand up to heavy traffic. This is basic technology but is often neglected by tile layers. The aim should be to keep the free water/cement ratio at or below 0·4, but at the same time to provide sufficient workability for thorough compaction by the means which will be used on site; this may well require the addition of a plasticiser to the mix. As compaction is fundamental to the long term durability of the floor, it should be checked as the work proceeds and prior to the laying of the tiles on the bedding mortar.

For large jobs, laboratory tests to determine the optimum density of the specified mix should be carried out, and these should be followed by site trials to determine what can be obtained under working conditions. For example, if, in the laboratory, a bulk density of about $2200\,kg/m^3$ was obtained, then on site it should be possible to obtain a density in the range of 2000 to $2100\,kg/m^3$. Density checks on site can be made by hammering steel cylinders of known volume and weight into the plastic mortar as soon as compaction is finished. A simple calculation will yield the bulk density of the mortar.

When the Engineer or Clerk of Works is satisfied that the mortar bed has the specified density, tile laying can proceed. The above procedure envisages a very high standard of site supervision and workmanship, both of which are expensive. However, against this has to be considered the financial loss to the client and claims against the contractor if the floor fails after the warehouse, etc., is in operation.

Joint widths should be kept as narrow as possible, but account must be taken of the tolerances on the size of the tiles used. As the joints are likely to be the weakest part of the floor, minimum joint width should be taken into account when selecting the type of tile to be used. Generally, for tiles of low manufacturing tolerance, a joint width of 3–4 mm is adequate; for higher tolerances, it may be necessary to accept a width of 6–10 mm; the latter should be considered as a maximum, and care taken to ensure that it is not exceeded.

Movement joints should be located around the perimeter of all tiled areas, and directly over all movement joints in the structural slab. In suspended floors, it is advisable for movement joints to be provided along the centre line of supporting beams and other locations where there is negative bending (hogging moments). In ground floor slabs, movement joints should be provided directly over warping joints, as movement may take place along the line of such joints as moving loads pass along the floor. These movement joints should be for the full depth of the tiles and tile bed.

In the case of ceramic tiles, the general recommendation of manufacturers (which is repeated in the relevant Code of Practice and to which reference should be made) is that movement joints should be located in both directions at centres between 4·5 m and 9·0 m, and at the perimeter. It is advisable to follow the manufacturer's recommendations for all types of tiles. A spacing of 10·0–15·0 m in both directions is generally considered adequate for concrete and terrazzo tiles. Regarding jointing materials, joints between the tiles should be grouted in, the mix proportions depending on the width of the joint. For the narrow joints, up to 4 mm wide, a mix of 1 part cement to 1 part clean fine sand can be used. For wider joints, the mix should be leaner, say 1:3, as this will help to reduce drying shrinkage. The use of a plasticiser in the grout mix will also reduce shrinkage by reducing the amount of gauging water needed. The joints should be completely filled and compacted in one operation. Proprietary grouts can also be used; these often contain workability aids or water repellents.

The selection of suitable sealants for the movement joints can be difficult. Floors in warehouses, supermarkets and shopping centres are often cleaned daily with water and mechanical brushes. Unless the sealant is well bonded to the sides of the tiles, the cleaning process can pull out the sealant from the joint. The same type of failure can of course, also occur to the grouted joints.

Generally, thermal movement of the floor inside a building is likely to be small, and so a sealant possessing a relatively small extension but which will bond well to the sides of the tiles and which will be durable under working conditions is likely to be more suitable than softer, more elastic material. Insitu materials such as epoxies, epoxy/polysulphides, can be considered. It is advisable for a fairly hard back-up material (such as resin bonded cork) to be used in movement joints as it is this material which gives support to the joint sealant. In the opinion of the author this is better than using a softer material such as polyethylene strip as reliance then has to be placed on bond between the sealant and the sides of tiles to prevent downward movement of the sealant under traffic.

Joints are the weakest part of a tiled floor and more attention should be paid to them than is generally the case. The width of the joint in the structural frame should be maintained right through the floor finish; the use of a stainless steel cover strip, fixed on one side of the joint only, is recommended for heavily trafficked floors. It is essential that the sides of the joint through the floor finish should not break down under the action of traffic.

6.1.8 Resurfacing with Chemically Resistant Tiles and Bricks

This is a highly specialised field, and early consultation with firms experienced in the problems involved is essential. At the same time, the Engineer should have sufficient knowledge to be able to set out his basic requirements. Important factors are likely to include:

(a) The contractors being considered must have a proven record of successful work, which should be checked by the Engineer.

FIG. 6.5. Core through Nori paviors showing joint merely pointed in epoxy mortar instead of complete filling of joint.

(b) The use to which the floor will be put and the loading.
(c) It is not advisable to place complete reliance on the liquid-tightness of the joints between the tiles/bricks. Therefore, the bedding mortar must possess adequate resistance to the chemicals which will be in contact with the floor.
(d) Where spillage on the floor is by chemicals which are very aggressive to Portland cement concrete, then consideration should be given to the provision of a chemically resistant barrier over the

surface of the base concrete. The author favours insitu materials for this purpose rather than sheeting.

(e) The joints between the tiles/bricks must be completely filled with well compacted chemically resistant jointing material. Pointing is quite inadequate and should not be accepted (see Fig. 6.5).

It is important to remember that once the floor is laid, there is no possibility of checking whether aggressive chemicals are penetrating down to the base concrete; this is only discovered when failure occurs.

If at all possible, the new floor surface should be laid to a reasonable gradient to ensure the run-off of aggressive chemicals and wash water. A gradient of 1 in 60 to 1 in 90 would be adequate.

Special care is needed in detailing around floor gulleys and channels to ensure that the protection provided is complete as these are locations where trouble often occurs. The selection of the sealant for movement joints requires careful consideration. It should be accepted that periodic renewal will be necessary.

6.1.9 Resurfacing with Extra Heavy Duty PVC Tiles

When a change of use occurs to a building, it often happens that a different type of floor finish is required to meet the new operating conditions. This may mean a seamless, dustless and abrasion resistant surface; some examples are the pharmaceutical industry, printing, electronics, clothing and food manufacture, etc.

Floor toppings based on cement are not truly dustless, even though with a concrete topping of high abrasion resistance the amount of dust may be very small.

Ceramic tiled floors with narrow (2 mm) joints may not meet fully the requirement for a dustproof surface.

In cases where complete absence of dust is not essential, there is often the need to have the floor resurfaced in a very short space of time. This may prohibit the use of 'wet' trades such as concrete and mortar, ceramic tiles, etc. In both such cases, the selection of high quality polyvinyl chloride (PVC) tiles reinforced with silica quartz may provide the answer. The tiles are usually 300 mm × 300 mm or 600 mm × 600 mm, with a standard thickness of 3·2 mm. The tiles have to be laid on a clean dry subfloor with a smooth surface (see Fig. 6.6).

The PVC tiles themselves have good impact resistance, but they do not act as a cushion or impact absorber, and so the screed or concrete on which they are laid must be able to resist impact from moving loads, etc.,

FIG. 6.6. Floor finished with extra heavy duty PVC tiles (courtesy : Altro Ltd).

on the floor. If a cement mortar screed is to form the base for the PVC tiles, then this must be carefully specified to ensure adequate strength. The author recommends a mix of 1:3 by weight, a water/cement ratio not exceeding 0·4, the incorporation of either an SBR emulsion (10 litres to 50 kg cement) or silica fume (10% by weight of cement) and a plasticiser or superplasticiser determined by trial mixes. All this may sound very elaborate, but the failure of such a floor in use can be disastrous to all concerned.

This type of PVC tile is laid with special adhesives which depend on the moisture condition of the subfloor, and the adhesives are normally supplied by the tile manufacturers. The directions of the manufacturer must be strictly followed and it is always advisable to appoint a specialist contractor recommended by the tile maker. Figure 6.6 shows such a floor.

6.1.10 Resurfacing with Insitu Bonded Semi-flexible Toppings

This type of topping consists essentially of bituminous emulsions, Portland cement, good quality sand and high quality crushed rock aggregate. Such toppings are suitable for providing a new hard wearing surface to existing concrete floor slabs for a range of industrial and commercial uses. It is not suitable for use where wet and abrasive conditions prevail, nor where there will be spillage of oil, grease or solvents.

The following information has been largely obtained from Colas Products Ltd (formerly Shell Composites) with respect to their Mastic Flooring. The surface is virtually dustless with joints which are largely self-sealing, and it should have a useful life of up to 20 years, provided reasonable maintenance is carried out. It is the author's experience that no floor subject to heavy wear is maintenance free, and that as defects appear they should be repaired as soon as possible.

The topping can be laid to a minimum thickness of 12 mm, but if for any reason a greater thickness than 20 mm is required, the topping should be laid in two layers. Before the topping is laid, the base concrete should be cleaned and brought to a reasonable state of repair, including joints and cracks, by the methods recommended in Sections 6.1.1–6.1.3. After the completion of the preparation of the concrete base slab, the suction is 'killed' by damping down overnight, then a priming coat is applied (about 50% emulsion and 50% cold water). This is allowed to dry and then a bond coat of undiluted emulsion is applied, followed immediately by the mastic mix.

The topping is laid in alternate bays, about 2·0 m wide between metal side forms. Initial compaction is by a vibrating beam followed by a wood float finish. The trowelled surface is then rolled with a 200–250 kg roller. On completion of the rolling, the topping must be moist cured for two days (the curing is effected by the same methods as used for concrete).

6.1.11 Resurfacing and Repair with Cast Iron or Steel Grids

These two materials (cast iron grids and steel grids) are made in a limited range of patterns, and are suitable for floors subjected to very heavy moving loads, impact and abrasion. Cast iron is much more corrosion resistant than steel. The brief information given below relates to the

Metaplate (cast iron) and the Avonbath (steel) plates marketed by Prodorite Ltd.

The cast iron plates are made in two basic patterns, one being virtually closed on the surface and the other an open grid pattern. There are differences in the recommended methods of laying, but both are bedded in high strength fine aggregate (10 mm) concrete. The mix used should be similar to that used for a granolithic topping. The writer also recommends that if there is spillage of aggressive liquids onto the floor, as in dairies, then the concrete mix should be modified by the addition of 10–15 litres of SBR emulsion to 50 kg cement.

If the work is carried out under emergency conditions, then high alumina cement may be used as this will give high strength at 24 h (see Chapter 1).

The plates should be laid with plain butt joints. Complete details for laying the plates are provided by the suppliers. The steel (Avonbath) plates should not be used where they will be in contact with aggressive liquids. They are bedded on fine concrete as are the cast iron grids, with plain butt joints. They give a smooth, almost jointless floor finish and are used in large bakeries where the floors are subjected to heavy wear and impact.

6.1.12 Resurfacing to Resist Attack by Ammonium-based Chemical Fertilisers

It was stated in Chapter 2, Section 2.2.3, that ammonium-based fertilisers can be very aggressive to Portland cement concrete; this is due to the presence mainly of ammonium nitrate and sometimes of ammonium sulphate. One solution to this problem is to provide an inert barrier over the whole of the floor area on which the fertilisers are stored. Such a barrier would have to be abrasion and impact resistant as well as inert to chemical attack; suitable toppings are very expensive, particularly as the areas to be covered are likely to be large.

It has been found practical and economical to use a high quality concrete containing an admixture which increases the resistance of the concrete to chemical attack and this increases the useful life of the floor. The admixture with which the author has had experience is Corrocem, made in Norway by NORCEM, the Norwegian cement manufacturer. Corrocem contains a superplasticiser and a high proportion of refined silica fume (silicon dioxide). It is a very fine powder of non-uniform particle size and should be used with Portland cement having a low C_3A content (not exceeding 5%). The recommended proportion of Corrocem is 20% by weight of cement, and the minimum cement content of the mix is 360 kg/m^3, with a maximum water/cement ratio of 0·40. Special care is required in batching, mixing and curing. Detailed directions for site use are issued by the manufacturers and their agents in various countries, as Corrocem is used on a world-wide basis.

In addition to increasing resistance to chemical attack by ammonium nitrate and other chemicals, the addition of Corrocem will increase strength at early ages and at 28 days by up to 50% compared with a control mix, and permeability is substantially reduced. Tests have shown that the electrical resistance of the concrete is increased by about 200% and this is claimed to help reduce the risk of rebar corrosion.

6.1.13 Tolerances of Surface Finish

In recent years there has been a significant increase in the demand for floor surfaces which have very low tolerances on line and level. These 'superflat' floors are needed for the satisfactory operation of high rise turret type trucks which run in the traffic ways between high racking.

While in some cases the changeover involves a completely new structural floor slab, there are many instances where all that is needed is new, high quality, thin-bonded topping which can be laid to the strict tolerances specified.

It is the production of a comprehensible and practical specification for such a 'superflat' floor which gives rise to considerable difficulty, and may end in litigation between the client, the flooring contractor and the specifier. The basic recommendations for tolerances of finishes to concrete floors are contained in clause 120 of the Code of Practice for Insitu Floor Finishes (CP 204). This recommends ± 3 mm under a 3·00 m straight edge, measured as directed in the Code for localised variations in level, and ± 15 mm for large areas. The author has found that there are often serious misunderstandings of the terms 'level' and 'flat' and 'gradient' or 'slope', and Fig. 6.7 illustrates the essential differences between these terms.

In the US the E. W. Face Company have been one of the pioneers in the field of specifying and constructing superflat floors, while in the UK, it has been the IDC Group PLC. Both companies have prepared technical literature and contributed articles to construction journals on the subject. In the UK the Cement & Concrete Association developed a floor profilometer which records on a chart the true profile of the strip of floor over which it passes.

When close tolerances have to be specified and adhered to, to enable

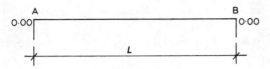

The floor between A and B is level and flat

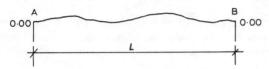

The floor between A and B is level but not flat

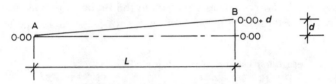

The floor between A and B is flat but not level,
the gradient is : *d* in *L*

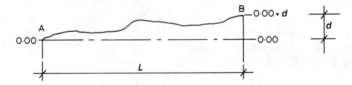

The floor between A and B is not level and not flat
the gradient is : *d* in *L*

FIG. 6.7. Illustration of terms 'Level', 'Flat' and 'Gradient'.

high turret trucks to run safely over the floor, great care must be taken in the wording of the contract; this is best done by close co-operation between the supplier of the trucks, the specifying engineer, the client, and the contractor who will have to lay the floor topping.

6.1.14 The Problem of Slippery Floors
High quality, abrasion resistant concrete floors have an acceptable

coefficient of friction and can be considered as slip resistant in their original state.

However, in use, two changes can take place which will reduce the slip resistance. These are:

(a) A general polishing of the surface under the action of traffic.
(b) Contamination of the surface with very fine particles of materials arising from the use to which the floor is put.

The experience of the author in investigating complaints of slippery floors is that the cause is generally a combination of both the above points. In addition there is the very important factor of the type of footwear used by employees working on the floor.

A considerable amount of research has been carried out into the characteristics of the floor surface ((a) and (b) above) and materials used for footwear. A few references are given in the Bibliography at the end of this Chapter. Organisations with special knowledge and experience in this field are:

British Ceramic Research Association (BCRA)
Rubber & Plastics Research Association (RAPRA)
Shoe and Allied Trades Research Association (SATRA)

There is a conflict between the need for a floor surface which can be easily cleaned and does not 'collect' dirt, and one which has a high coefficient of friction. The latter depends on the friction characteristics of the floor surface and the footwear.

The British Ceramic Research Association developed a piece of equipment for measuring the coefficient of friction of any type of floor surface *in situ*; this is known as the TORTUS and it can be purchased or hired. The equipment gives a direct visual reading of the coefficient of friction of the floor surface measured by a slider held in contact with the floor under a uniform load. The slider used can be in rubber, leather or other materials (see Figure 6.8).

In any investigation of alleged slipperiness it is necessary to establish what is causing the low coefficient of friction. Also to ascertain, by means of the TORTUS, what the coefficient of friction really is.

While the author feels that for general use on factory floors the TORTUS is the most suitable equipment, there is also the equipment developed by the Transport and Road Research Laboratory (TRRL) for measuring skid resistance. The two methods are compared in a Report by RAPRA in 1984 (see Bibliography).

FIG. 6.8. View of the TORTUS floor friction tester (courtesy: British Ceramic Research Association).

When concrete floors are subjected to heavy traffic, the type of aggregate used can have a significant effect on the slip resistance. This is why limestones are not recommended as aggregate in the concrete for motorways (in the surface layer) as these polish more than flint gravels and igneous rocks.

The remedy for a slippery factory floor will depend on a diagnosis of the cause and careful consideration of practical measures to improve the situation. Cleaning methods should be evaluated. The factory process cannot be changed. Advice can be obtained from SATRA on suitable footwear for the workers.

Obvious measures include roughening the surface of a concrete floor by acid etching or scabbling, or the provision of a complete non-slip coating consisting of a polymer resin and a special fine aggregate. Such coatings are very expensive and the floor surface must be carefully prepared so that all contamination is removed prior to application. Unless the cleaning method is satisfactory, the improvement may be only temporary. Such a high friction surface is also more difficult to clean.

6.2 REPAIRS TO CONCRETE ROOFS

6.2.1 General Principles, Including Tracing Leaks

The main reason for repairing a concrete roof is to remedy leakage. The problems involved are complex and the finding of a satisfactory solution requires considerable experience. In addition, the cost of the repair is usually high.

To identify the location of the leak or leaks can be very difficult as most roofs are provided with a waterproof layer on the top surface and defects in this layer very seldom coincide with the place where the water enters the building on the underside of the slab. In some cases the waterproof membrane is located below the thermal insulation (the 'upside down' roof), and this increases the difficulty in finding the leak. In addition to the waterproof layer (membrane), most roofs of buildings, except multi-storey car parks, are provided with thermal insulation and vapour barrier and often with a suspended ceiling as well.

Flat concrete roofs in the UK are subjected to a wide temperature range; the actual range depends on whether the external surface is protected by thermal insulation, and also on the colour. BRE Digest 228, Table 2, gives the range for external light coloured materials in roofs and walls as $-20°C$ to $+50°C$. For dark coloured materials the range is increased to $-20°C$ to $+65°C$. It is obvious that the temperature range will be reduced if thermal insulation is placed on the top of the slab or the exposed finish is light coloured.

Roof slabs of multi-storey car parks are not usually provided with thermal insulation, but must be watertight because the concrete slab itself cannot be relied upon. Water which seeps through concrete becomes highly alkaline and if it drips onto cars parked below it will damage the paint. More detailed comments on repairs to roofs of multi-storey car parks are given in Section 6.2.2.

When leakage has occurred, the first step in the investigation is to locate the points of leakage, and the source of the water. The dampness showing in the building beneath the roof slab may be due to leakage through the roof membrane, water penetration which is by-passing the membrane due to the absence of, or serious defects in, a damp-proof course in the roof parapet; it may also be interstitial condensation, or surface condensation due to inadequate thermal insulation and/or in-adequate heating in the building. It is essential, but by no means easy, to establish as soon as possible in the investigation, which of these is the real cause of the trouble.

In this book only leakage through the roof slab will be considered, and readers should refer to the bibliography at the end of this chapter for publications on interstitial and surface condensation.

Between 1970 and 1974, the Building Research Establishment at Garston made detailed investigations of over 500 defective buildings, of which 254 were 'dampness' defects. Of these, 136 were rain penetration of which 49 were through roofs; 90 were condensation comprising 56 surface and 34 interstitial; and 19 arose from entrapped water in the roof structure or in insulation. Details are given in BRE Digest 176, 1975.

Unless it is clear from a careful visual inspection that the waterproof membrane has suffered general deterioration and should be renewed in whole or part, it will be necessary to locate the points of leakage.

Where a mastic asphalt membrane has deteriorated due to age it is often possible, and is certainly quicker and less disruptive, to apply a new polymer (spray or brush applied) membrane over the asphalt. Roofing felt can be rewaterproofed by 'torching' on a new high performance membrane. Some basic recommendations are given later in this Section.

The main sources of leakage are likely to be at the following locations which should be inspected first:

(a) The junction of the roof with vertical surfaces, e.g. parapet and other walls.
(b) At points where pipes and other services penetrate the roof.
(c) At movement joints.

While ponding always looks suspect, its presence is no real indication that leakage is occurring there.

When a careful visual examination on the roof has failed to locate the place(s) where the membrane is defective, and water penetration through

defects in parapet walls can be eliminated, then a detailed search has to
be made for the leak(s). A method which the author has found to give
reasonably satisfactory results, when practicable to use, is to divide the
roof into a number of small areas, each about 1·5 × 1·5 or 2·0 × 2·0 m,
These are then ponded to a depth of at least 100 mm, taking each area in
turn. A dye is added to the water in each ponded area. The dye must be
one which will not be decolorised by seepage through the membrane and
the highly alkaline concrete. Firms such as W. S. Simpson & Co. Ltd. of
London and CIBA-GEIGY Ltd of Macclesfield can supply suitable dyes.
This is a tedious and time consuming process, but the alternative to
locating the leak in the membrane and repairing it is to strip the roof and
lay a new membrane, or completely renovate as briefly described in
Sections 6.2.3–6.2.5.

A piece of equipment which has recently come onto the UK market is
an electrical conductance meter which when passed over the roof surface
shows on a display panel variations in electrical conductance of the
roofing material caused by variations in moisture content. The equipment is made in two models, one which is mounted on a frame with
wheels, and the other smaller unit which is hand held. Even the larger
one is readily transportable and operates from batteries; as the meter
passes over the roof surface the electrodes on its base record changes in
the electrical conductance of the roof.

6.2.2 Methods of Waterproofing

It is necessary for the consultant investigating defects in flat concrete
roofs to have reasonable knowledge of the various methods of flat roof
construction and the types of materials used for thermal insulation and
waterproofing.

The two basic methods of roof construction are:

(a) The 'standard' method with the waterproof membrane on the
    exposed surface of the roof and the thermal insulation below with
    a vapour barrier between the insulation and the concrete slab.
(b) The 'upside down' roof, where the waterproof membrane is below
    the thermal insulation which must then be impermeable to water.

The main materials used for the waterproof membrane are built-up
roofing felt and mastic asphalt; other materials which are sometimes used
are various types of polymer sheeting and *in situ* polymers applied by
brush or spray to the concrete or screed.

The sheeting materials are mainly PVC, butyl rubber, chlorinated polyethylene (CPE) and Hypalon, but there are others on the market. Many suppliers recommend that these sheeting materials be laid as an unbonded sheet except at the perimeter, but the author recommends strip bonding using a non-moisture sensitive adhesive. The bonded strips should be at about 1·0 m to 1·5 m centres in both directions. This strip bonding has two advantages; it helps to prevent creasing during laying, and if after laying it becomes damaged, water penetrating the defect is confined to the unbonded squares of about 1·0 × 1·0 m or 1·5 × 1·5 m. Unless there are leaks in the base concrete within these squares, water will not be able to enter the building. If it does, the location of the defect can be more easily found.

When new waterproof membranes have to be laid, it is most advisable that they should be of the highest quality because they have to remain watertight under very arduous climatic conditions for a long period of time.

While the suppliers of all types of materials for the waterproofing of flat concrete roofs provide recommendations for laying, the following basic principles should be observed:

Mastic asphalt. The concrete slab or screed should have a reasonably smooth surface on which the separating membrane has to be laid. Some authorities recommend a glass fibre tissue material as the isolating membrane in preference to bituminous sheathing felt.

The asphalt is in two grades, roofing grade and paving grade. For roofs subject only to occasional foot traffic for inspection and repair, the roofing grade can be used. It is laid to a total thickness of 25 mm in two layers, with the top coat having 5–10% additional grit.

The author recommends that on the completion of new asphalt work, the roof should be given one or more coats (as recommended by the paint manufacturer) of solar reflective paint. The paint used must be specifically formulated for compatibility with the asphalt (and the same applies to other roofing materials). High quality paints of this type will reduce the rate of deterioration caused by thermal-chemical reactions and by ultra-violet light. It will also reduce mechanical stress by reducing temperature rise on hot sunny days.

6.2.3 Renovating Mastic Asphalt
When the asphalt has, due to aging, lost some of its waterproofing properties, but is otherwise reasonably sound, it can be 'rejuvenated' by the application of a suitable polymer coating, about 1 mm thick. The basic requirements for such work are set out below:

(a) The asphalt must be carefully checked for physical defects which must be cut out and made good, usually with an epoxy mortar.
(b) The whole surface of the asphalt should be thoroughly cleaned by wire brush scrubbing with a detergent and then well washed down with clean water, or by grit blasting and cleaning.
(c) After the surface has dried, a primer is applied and allowed to cure.
(d) Following the primer, the selected polymer coating is applied to give a dry film thickness of not less than 1 mm.
(e) While the polymer coating is still tacky, a slip resistant fine aggregate is sprinkled on; a calcined bauxite would be suitable.
(f) Special attention must be paid to skirting and exposed edges, such as may occur on external balcony slabs.

FIG. 6.9. Resealing of built-up roofing felt by torching (courtesy: Ruberoid Ltd).

6.2.4 Renovating Built-up Roofing Felt

The author is indebted to Ruberoid Building Products Ltd for the information which follows:

(a) The existing roofing felt should be cleaned and major physical defects made good.
(b) On the prepared surface, 'Rubertorch HP400' (a high performance polyester based membrane) is bonded by 'torching' using special equipment; the flame is applied to the lower surface.

Figure 6.9 shows the method of application. It should be noted that the 'Rubertorch HP400' can also be used for renovating mastic asphalt.

6.2.5 Renovation by Application of Stainless Steel Sheeting

This system for complete weather-proofing of existing flat roofs where the waterproof membrane requires complete renewal, has recently been introduced onto the UK market. It consists of Type 316, austenitic stainless steep strip 0·4 mm thick and 650 mm wide. The strip is formed on site into trays with upstand flanges on the long edges to give 20 mm high seams at 600 mm centres. The seams are formed by welding and then folded over above the weld with a special machine. The steel strips are fixed to marine quality plywood or moisture-resistant chip-board. The system is designed, supplied and fixed by Broderick Structures Ltd of Woking to whom the author is indebted for the above brief description. A 10 year guarantee is given, but the system is only a practical proposition for roof areas in excess of 500 m².

6.3 ROOF GRADIENTS TO FACILITATE DRAINAGE

It is often found that gradients for drainage of the roof are inadequate. The Code of Practice, CP 144, recommends 1 in 80 as the minimum. Other authorities recommend an overall gradient of 1 in 40 so as to help prevent ponding. Once the roof has been constructed it is seldom that anything effective can be done to improve the gradient.

Movement joints in the roof structure can be vulnerable to water penetration and should be given detailed consideration when specifying remedial work. The flat roofs of most buildings are generally only accessible for the purpose of inspection and repair. In such cases, full movement joints (generally called expansion joints) should preferably be ɔf the double kerb type as this detail can be made completely watertight much more easily than the flush type of joint.

6.4 ROOFS OF MULTI-STOREY CAR PARKS

Mention was made in Section 6.2.1 of the need to provide a waterproof membrane on concrete roofs of car parks as the concrete cannot be relied upon to be completely watertight. It is the opinion of the author that such roofs should have the same standard of watertightness as the roof of a dwelling. The roofs of many multi-storey car parks are used for car parking and therefore the finished surface has to stand up to vehicular traffic (see Fig. 6.10).

FIG. 6.10. Waterproofing of car park roof deck with 'Tretodek' elastomer (courtesy: Tretol Building Products Ltd).

The two basic systems used for the running surface are mastic asphalt and proprietary spray/brush applied polymers. When the floor below the roof is used for parking, thermal insulation is usually omitted, and the waterproof membrane forms the wearing surface. The location of leaks, and methods of repair are similar to those discussed in Sections 6.2.1–6.2.4. It must be kept in mind that the roof slabs of multi-storey car parks are likely to be subjected to a wider temperature range than in buildings where the lower floors are heated and thermal insulation provided.

The design recommendations for multi-storey car parks issued by the Institution of Structural Engineers in 1976, indicate that bonded membranes should be capable of bridging a 0·3 mm wide crack in the concrete base slab. The Department of Transport require that for bridge decks, the waterproof membrane must be capable of bridging a 0·6 mm wide crack in the base concrete. At the time of writing this book, the British Board of Agrément had issued eleven Certificates for membranes for bridge decks. Probably the area which gives the greatest trouble is the sealing of the full movement joints (expansion joints). These joints must be 'flush' and as previously mentioned this type of joint is much more difficult to seal than the raised kerb type.

It is now generally agreed that for flat roofs subject to vehicular traffic, a mechanical joint such as the Radflex is the one most likely to prove satisfactory in the long term (see Fig. 6.11). Kimber, in his excellent paper

FIG. 6.11. Movement joint in car park roof slab completed and sealed with Radflex 5200 (courtesy: Radmat (London) Ltd).

on the investigation of defects in multi-storey car parks, says ... 'no one can yet say with any certainty how any given expansion joint is going to behave ... with the most skilful design of a building it seems impossible to determine how that building will move.'

Where the roof is inverted (upside down roof), with the thermal insulation above the waterproof membrane, a special running surface must be provided. Such a system using precast concrete flags has been developed by the Cement & Concrete Association and reported on in a paper by R. G. Chaplin in 1982.

6.5 ROOF GARDENS

There has been an increasing interest in recent years in the provision of roof gardens which often incorporate what are known as 'water effects'; these water effects consist of fish and lily ponds, and waterfalls. Roof gardens with water effects pose special problems for the designer and the contractor, and even more for the consultant who may be called in to advise on how to remedy leakage and rectify other defects which are mentioned below.

Remedial work may entail the provision of completely new waterproofing, but even this may not effect a complete cure as the new membrane has to be applied to the existing structure. A fundamental requirement is that waterproofing materials must not be toxic to plant and/or fish life.

Materials based on Portland cement will leach lime into the pool water when new and this will increase the alkalinity (raise the pH) substantially. Therefore pools finished with cement/sand rendering must be allowed to mature by being filled and emptied several times over a period of about a month and walls and floor washed down between each filling.

For polymer based waterproofing membranes, the manufacturers' certificate of non-toxicity should be obtained. Any material approved for lining drinking water reservoirs would be suitable, e.g. Colebrand's CXL230.

While a rigid cement based rendering is sometimes used for waterproofing, the author prefers a brush or spray applied polymer which possesses some degree of flexibility and can bridge cracks up to about 0·3 mm.

Waterproofing work is highly specialised and is expensive, but it should only be entrusted to contractors with a proven record of successful work.

A problem which sometimes arises with roof gardens is that the original design allowed drainage to take place partly through cement/ sand screed discharging to manholes which in turn discharged to rain water pipes. Water passing through a fairly porous cement based mortar or concrete will leach out calcium carbonate, which in contact with the carbon dioxide in the air is converted to calcium carbonate. In two cases investigated by the author, 100 mm diameter pipes were more than half filled with a deposit of calcium carbonate. The original intention had not been for the water to drain through the screed, but as the screed had not been provided with a waterproof membrane on the top, a high percentage of the drainage water had seeped through the screed.

The better solution was to remove all the garden soil and lay a membrane on the screed. This was not acceptable, and so a recommendation was made to seal off the edge of the screeds which were exposed in the drainage manholes and to employ a special firm to remove the heavy deposits of calcium carbonate in the rain water pipes.

BIBLIOGRAPHY

AMERICAN CONCRETE INSTITUTE, *Guide for concrete floor and slab construction,* ACI Committee 302; ref. 302.1R-80, 1980, p. 46.

BRITISH STANDARDS INSTITUTION, *Code of Practice for flat roofs with continuously supported coverings,* BS 6229, 1982, BSI, London.

BRITISH STANDARDS INSTITUTION, *The structural use of concrete,* BS 8110, Parts 1 and 2; 1985 (replaces CP 110).

BRITISH STANDARDS INSTITUTION, *Roof coverings,* CP 144, Parts 3 and 4, BSI, London.

BRITISH STANDARDS INSTITUTION, *Insitu concrete floor finishes,* CP 204, BSI, London.

BRITISH STANDARDS INSTITUTION, BS 5395, Part 1—Stairs, BSI, London.

BUILDING RESEARCH ESTABLISHMENT, *Corrosion resistant floors in industrial buildings,* Digest 120, Aug. 1970, p. 8.

BUILDING RESEARCH ESTABLISHMENT, *Estimation of thermal and moisture movements and stresses,* Parts 1, 2 and 3; Digests 227, 228 and 229, July to Sept. 1979.

BUILDING RESEARCH ESTABLISHMENT, *Solar reflective paints,* IP/26/81; Dec. 1981, p. 4.

CHANDLER, J. W. E., *Design of floors on the ground,* Technical Report 550, Cement & Concrete Association, ref. 42.550, June 1982, p. 22.

CHAPLIN, R. G., *The regularity of concrete floor surfaces; a survey of current knowledge*, CIRIA Report No. 48, Jan. 1974, p. 31.

CHAPLIN, R., *Abrasion resistant concrete floors*, Paper at International Conference on Concrete Slabs: Materials, Design, Construction and Finishing, Dundee, April 1979, p. 14.

CHAPLIN, R. G., *Development of a roof cladding system for car parks and light traffic*, Paper at Concrete International, Brighton, April 1982, p. 11.

DEACON, R. C., *Concrete ground floors*, Cement & Concrete Association, ref. 48.034, 1982, p. 30.

DORE, E., Drainage, waterproofing, floor finishes and maintenance of multi-storey car parks. Paper in Session 2 of the *Joint Conference of the Inst. Struct. Eng. and Inst. H. Eng.*, May 1973, pp. 47–57.

FREY, G., Construction considerations for serviceable concrete floors. *ACI Journal*, June 1973, title 70–42; ACI Committee 302, 416–34.

HARRISON, R. AND MALKIN, F., On-site testing of shoe and floor combinations. *Ergonomics*, **26** (1), 1983, 101–8.

JAMES, D. I., *Slip resistance tests for flooring: two methods compared*, Rubber & Plastics Research Assoc. (RAPRA), Members Report no. 94, 1984, p. 16.

KIMBER, G., *Investigations of defects in multi-storey car parks*, Paper at meeting of the British Parking Association, Feb. 1979, p. 7.

LIU, T. C., Abrasion resistance of concrete. *ACI Journal*, title 78–29, Sept.–Oct. 1981, 341–50.

MALKIN, F. AND HARRISON, R., A small mobile apparatus for measuring the coefficient of friction of floors. *Journal of Physics D, Applied Physics*, **13** (1980), 77–9.

MINISTRY OF PUBLIC BUILDING AND WORKS, *Condensation*, Part 1—*Design Guide*, 1970, and Part 2—*Remedial Measures*, 1971, HMSO, London.

PACKARD, R. G., *Slab thickness design for industrial concrete floors on grade*, Portland Cement Association, USA, 1976, p. 16.

PATEMAN, J., Specifying and constructing flat floors. *Concrete*, **18** (3), March 1984, 7–9.

PERKINS, P. H., Membranes in floor construction. *Journal of Concrete*, January 1983, 23, 24.

PERKINS, P. H., *Floors, construction and finishes*, Cement & Concrete Association, Viewpoint publication 12.057, 1973, p. 128.

RINGO, B. C., Design, construction and performance of slabs-on-grade for industry. *ACI Journal*, Nov. 1978, title 76–61, 594–602.

SPEARS, R. E., *Concrete floors on ground*, Portland Cement Association, USA, 1978, ref. EB/075.01D, p. 29.

THE EDWARD W. FACE CO., *Specification of floor flatness*, Technical Sheet 831, 1983.

WARLOE, W. J. AND PYE, P. W., *Floor screeds, bakery floors, synthetic resin flooring*, Building Research Establishment, CP 28/74, 1970, p. 14.

Chapter 7

The Repair of External Wall Tiling and Mosaics, Brick Slips and Rendering

7.1 GENERAL CONSIDERATIONS

During the past 25 years there has been considerable use of ceramic tiles and mosaic as an external finish to buildings, mainly multi-storey. On the external exposed surfaces of beams and projecting floor slabs, brick slips are often used to hide or disguise the concrete. However, in many cases, considerable trouble has arisen due to the failure of the tiles, mosaic and brick slips and rendering to remain securely bonded to the concrete background. These failures are not confined to tiles and mosaic fixed on site, but have also occurred to factory made precast concrete elements where the tiles/mosaic were fixed in the factory as the units were cast.

7.2 EXTERNAL TILING AND MOSAIC

The use of tiles and mosaic has the great advantage that the surface is virtually self-cleansing and therefore weather staining is reduced to a minimum.

For the purpose of this section, tiles and mosaic are considered as one material and will generally be referred to as tiles only.

Due to inaccuracies in the surface of the concrete, it is usual practice to 'dub-out', apply one or more coats of rendering, and then bed the tiles in either a cement/sand mortar or a proprietary adhesive. The best practice is to design, specify and construct the concrete so that the achieved tolerances on the surface finish which has to be tiled are sufficiently close that the tiles can be fixed direct to the concrete without any dubbing-out or rendering. However, this is difficult to carry out in practice.

Tiles and mosaic used externally on buildings should be frost proof. The tiles vary in thickness from about 10 to 15 mm, and the mosaic is about 6 mm. The thickness of the bed depends largely on whether a cement/sand mortar or a cement based adhesive is used. When a cement based adhesive is used the thickness would be about 3 mm. The thicker tiles made in the UK have deep recesses on the back, while the thinner tiles (mostly imported from Germany) have shallower recesses.

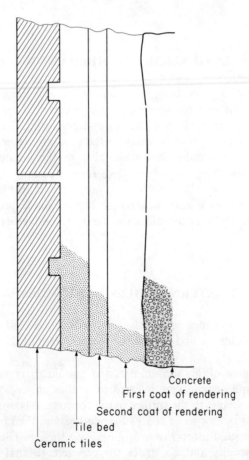

Concrete
First coat of rendering
Second coat of rendering
Tile bed
Ceramic tiles

Fɪɢ. 7.1. Diagram showing tiling and rendering on concrete substrate.

7.2.1 Reasons for the Failures

As stated above, the 'failure' of external tiling is invariably due to loss of bond at one or more levels in the construction. This loss of bond can be slight, medium or severe. In the latter case tiles can fall from the building.

Figure 7.1 shows in diagram form a section through a tiled concrete member as the work is normally executed. It should be noted that if the concrete surface is very irregular, dubbing-out would be needed in addition to the rendering, thus creating an additional layer in various places.

From the diagram it can be seen that there are four planes or interfaces, at each of which loss of bond can occur. For practical purposes, the range of thickness, measured from the face of the concrete to the face of the tiles/mosaic, is likely to be as given below:

Tiles		Mosaic	
Two-coat rendering:	15 mm	One-coat rendering:	10 mm
Mortar bed:	10 mm	Bed (adhesive):	3 mm
Tiles:	10 mm	Mosaic:	6 mm
	Total: 35 mm		Total: 19 mm

In the two examples given above, the vertical loading is approximately 42 kg/m² for the mosaic and 78 kg/m² for the tiling. Loading of this magnitude will result in an appreciable shear stress at the interface between the rendering and the concrete substrate, and also to a lesser degree between the various layers.

As everything generally appears to be in order when the job is finished, what occurs subsequently to cause reduction in bond to an extent which results in dislodgement of the tiles? There are many reasons for this loss or reduction in bond, and it is usually difficult to decide with certainty which ones apply in a particular case, and even more difficult to define their relevant importance. It is invariably a combination of factors which eventually leads to the failure.

The principal causes of bond failure may be summarised as follows:

1. Interface: concrete to rendering:

    (a) Inadequate preparation of the concrete; for good bond a sound mechanical key is necessary and this means exposure of the coarse aggregate by bush hammering, grit blasting or high velocity water jets. The prepared surface must be free from dust and grit and should be damp.

(b) Poor quality weak concrete of inadequate strength to hold the weight of rendering and tiles.

(c) The application of the first coat of rendering to too great a thickness; the maximum recommended is 12 mm.

(d) Insufficient compactive effort when applying the rendering.

(e) Poor grading of the sand for rendering; also the presence of impurities such as silt, clay and organic matter. The presence of silt and clay particles will show up in the grading, but unfortunately, no specific limits can be set for organic contamination.

(f) Inadequate protection and curing of the rendering. The rendering must be protected against strong winds, hot sun, heavy rain and frost, and must not be allowed to dry out too quickly.

(g) Corrosion of reinforcement in the concrete resulting in cracking and spalling of the concrete after the completion of the tiling.

2. Interfaces: first to second coat of rendering and rendering to bedding:

(a) Lack of an adequate key due to omission of deep combing of the surface of the rendering.

(b) The application of a rich mix onto a leaner mix below. Each coat, including the bedding mortar, should be no richer than the preceding one. The second coat of rendering should also be thinner than the first coat.

(d), (e) and (f) as above.

3. Interface: bedding to tiles:

(a) Failure to ensure full bedding of the tiles which includes complete filling of the 'frogs' at the back of the tiles.

(b) The use of too thick a bed.

(c) The use of too rich a mix for the bedding mortar.

4. General:

Inadequate provision for movement both horizontally and vertically of the various layers including the tiling, when these are considered as a monolithic unit. The concrete which forms the base to the various layers will deform under load, and deformation will occur as deflexion, shrinkage and creep. In addition, the tiles and supporting layers will be subjected to considerable thermal movement, both short term and long term. The concrete structure will be subjected to thermal movement of a different magnitude and a different time scale. The colour of the tiles can have an appreciable effect on the surface temperature reached on hot sunny days, also the orientation, whether north, south, etc. Stress will therefore develop in the various layers due to the thermal gradient from outside to inside. Water penetration can have disruptive effects when it freezes. Horizontal surfaces are much more vulnerable than vertical ones, as the joints between the tiles, as they are normally made, cannot be relied upon to be watertight.

All joints should be finished flush with the face of the tiles.

7.2.2 Investigations Required to Assess Remedial Work Needed

While an examination of the original specification and drawings can be useful, it should not be relied upon to give factual information on what was actually done on site. There is no substitute for careful site investigation, and the following procedure is recommended:

(a) Small areas of tiling should be carefully removed, including the bedding and rendering, back to the base concrete. As each layer is removed it should be recorded and preserved in its original state for detailed examination off the site. In addition it may be prudent to carry out a limited amount of coring in those areas which show little or no signs of distress.

(b) The whole surface of the tiling should be carefully examined visually, and by tapping, to locate any cracks and hollow-sounding (debonded) areas.

(c) The position of any movement joints should be noted as well as the condition of the sealant used, and its type.

(d) The samples of bedding mortar/adhesive and rendering should be of adequate size to enable a thorough examination, supplemented when necessary by analysis, to be carried out. The object of the off-site examination of the samples is to provide additional information so that the reasons for the failure can be established with confidence.

(e) If on removal of the tiles and rendering it is found that the concrete is cracked and/or spalling, then this must also be investigated by the methods described earlier in this book.

The investigation outlined above will be expensive as cradles or scaffolding will be required (see Fig. 7.2).

FIG. 7.2. Investigation from cradles of tiled face of building (courtesy: Lionel Arnold (Tile Fixers) Ltd).

7.2.3 Remedial Work

Each job has to be dealt with as a separate problem, because that in fact is what it is. The causes of failure and methods of repair will vary from site to site even though the basic reasons for loss of bond and the principles of repair remain substantially the same.

The most difficult problem which has to be solved realistically relates to the extent of the remedial work, that is how much tiling should be completely removed and how much can remain. The most simple, but most expensive, is to remove the whole of the tiling which exhibits any sign of loss of bond, right back to the base concrete, and start again from there. Recommendations for this new work are not given here as they are described in detail in the latest revision of British Standard BS 5385:Part 2:1978—External Ceramic Wall Tiling and Mosaics, and BS 5262—External Rendered Finishes.

Recommended methods of repair, involving only a limited amount of removal, and preserving the existing tiling as far as possible, are described below.

    1.   All tiles, rendering, etc., which exhibit significant loss of bond and

bulging should be removed back to a sound, well bonded substrate, which may require going back to the base concrete. There is no reason to remove large areas of sound well bonded rendering where the debonding has occurred only between the rendering and the tile bedding. Considerable experience is required to decide correctly what has to be removed and what can be safely left. While the final decisions may have to be taken on site for some areas, the original investigation and testing outlined above should form the basis for such decisions. The use of percussion tools should be avoided as these set up considerable vibration which can disturb otherwise sound areas. Cutting, etc., can be done by saws and in certain cases with high velocity water jets.

2. If the rendering is sound and only the tiles have debonded, then an assessment has to be made of the number of tiles to be removed and replaced. The author considers that if it is assessed that more than about 15% of the area of a tile has lost bond, then it would be prudent to remove it.

3. In a few cases it may be found that the tiles were fixed direct to the concrete, in which case the surface of the concrete must be properly prepared to ensure maximum possible bond with the tile bed. This preparation can be difficult because standard methods involving percussion tools (bush hammering, etc.) can weaken the bond to adjacent areas of tiling (see Fig. 7.3).

4. In executing the work, care must be taken to prevent water getting down behind existing tiles; if it remains there and subsequently freezes it can force the tile off the wall. It is therefore advisable for the tile edges and the perimeter of all areas from which tiling has been removed to be sealed with a polymer modified mortar; an SBR or proprietary mortar would be suitable.

5. The tiles can be refixed using a proprietary bedding mortar approved by the tile supplier.

6. All mortar joints, in both existing and new work, should be finished flush with the surface of the tiles; recessed joints should not be permitted as these are unlikely to be adequately weathertight.

7. Attention must be paid to the provision of vertical and horizontal movement joints, and this is discussed later in this Section.

8. In some cases it is found that the tiles are fixed to rendering and the rendering is in places laid on dubbing-out due to initial irregularities in the concrete. Such a situation can cause a lot of

FIG. 7.3. Defective concrete behind partly bonded tiles (courtesy: Lionel Arnold (Tile Fixers) Ltd).

problems, and may result in a decision having to be taken to remove large areas of tiling, rendering and dubbing-out.

9. Mention has been made of the weight of rendering, mortar bed and tiling (about 78 kg/m²). The Code of Practice for external tiling and mosaics, BS 5385, Part 2, 1978, indicates that it is prudent to consider the use of reinforcement in the rendering 'to ensure that, in the event of adhesion failure between the rendering and the background, the rendering and applied cladding remain intact and fully supported'. Clause 19.2 of BS 5385, Part 2, describes how this should be done. The author considers that this method should be adopted where any extensive area of tiling and rendering has to be removed and replaced.

10. Dubbing-out should be carried out as for repair to defective concrete and is described in some detail in Chapter 4. The galvanised welded fabric or stainless steel mesh must be securely pinned back into the base concrete after the dubbing-out has been completed.

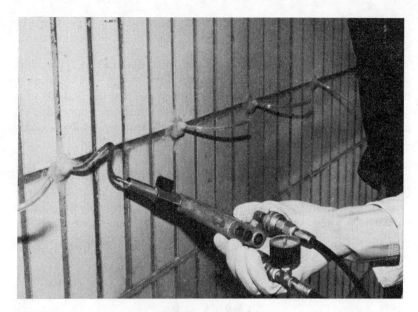

FIG. 7.4. Low pressure resin injection of delaminated external tiling (courtesy: Cementation Research Ltd).

In some cases it may be appropriate to improve the bond between the tiles and the substrate to which they are attached by the injection of a polymer resin at the interface. This technique has been successful in a number of cases; it requires considerable experience and highly skilled execution (see Fig. 7.4).

7.2.4 The Provision of Movement Joints in the Tiling

This subject is dealt with comprehensively in clause 20 of BS 5385:Part 2. In many cases of failure of tiling arising from debonding, inadequate provision for movement has been found to have played a major part. Movement joints in the tiling must be provided:

(a) Over existing movement joints in the background structure.
(b) Where the background materials change, e.g. from concrete blocks to insitu concrete.
(c) Where the tiling abuts other materials.
(d) Horizontally, at storey heights.
(e) Vertically, at not more than 4·5 m centres.

(f) At corners in the building structure; the movement joint should be located within 0·25–1·00 m of the corner.

The joint width should be in the range of 6–10 mm, and the sealant should be carefully selected for durability. The amount of movement to be normally accommodated is usually quite small, and the sealants used need have only limited flexibility. The author considers it advisable for movement joints to be carried through to the structural substrate and not just through the tiling.

It is particularly important in horizontal movement joints that the sealant should form a watertight joint and to do this it must bond well to the sides of the tiles. The use of a primer is generally necessary but in practice is seldom used. While polysulphides are probably the sealant most widely used, the author's experience is that their performance is unreliable. Flexible epoxies and epoxy-polysulphides are likely to prove more durable.

7.3 BRICK SLIPS

Brick slips are relatively thin pieces of clay brick which are used to mask the outer face of external reinforced concrete beams and exposed edges of floor and roof slabs. In some cases, where the concrete has been accurately cast to strict tolerances, the brick slips are fixed direct to the concrete. Generally however, the concrete requires rendering and frequently dubbing-out as well. The principles relating to bond and shear and shrinkage stresses are similar to those for external tiling; the difference lies in the fact that brick slips are fixed in long narrow horizontal bands.

Up to about the end of the 1970s it was usual to rely on adhesion (bond) between the brick slips and the substrate to ensure that the bricks did not slip. Unfortunately, this reliance was found in practice to be misplaced, as the high standard of workmanship required was seldom achieved. It is now prudent to insist on physical support in the form of either a single stainless steel angle for not more than three courses of bricks, or where the brick slips exceed three courses, special stainless steel units which support each course of bricks and are securely fixed to the structural concrete. These specials are made by such firms as Harris & Edgar Ltd at Croydon, George Clark of Sheffield, and Stainlessfix Ltd of Farnborough, Hants.

A properly sealed compression joint should be provided between the

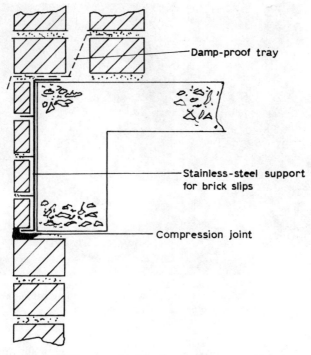

Fig. 7.5. Diagram of brick slips fixed to concrete beam.

bottom course of brick slips and cladding below. Where the panel walls are of cavity construction, the cavity tray at the bottom of the cavity must be detailed so as to project beyond the face of the brick slips below. This is shown in diagram form in Fig. 7.5.

The method of investigation where the brick slips have slipped is basically the same as for loss of bond in tiling and mosaics. Figure 7.6 shows three courses of slip bricks 'fixed' to the face of an edge beam; the absence of good bond can be clearly seen. In the case in question, the vertical movement joints were at about 8·0 m centres instead of a maximum of 4·5 m, and in some places the thickness of the dubbing-out and mortar backing exceeded 40 mm.

7.4 EXTERNAL RENDERING

Generally, external rendering on concrete beams and other members, where the rendering forms the final finish, consists of a cement/sand first

FIG. 7.6. View of debonded brick slips showing thick bedding behind and gap at concrete/mortar bedding interface.

coat (applied after any necessary dubbing out), and a finishing coat containing special aggregate to provide a pleasant and attractive finish. Failure of the rendering, except when due to sulphate attack, invariably results from high drying shrinkage stresses which cause cracking and debonding.

The principles of good practice in the application of external rendering are well known and publications giving detailed recommendations are readily available. Unfortunately, these recommendations are not always adhered to in specifications nor in the execution on site.

The two main publications dealing with external rendering are the Code of Practice, BS 5262, and the booklet issued by the Cement & Concrete Association, and readers should refer to them for detailed information.

As far as remedial work is concerned, the following summarises the steps to be taken, which are based on the assumption that the rendering has been applied to Portland cement concrete:

(a) The debonded areas should be identified by tapping with a light hammer and preferably marked. Usually, but not always, these areas are associated with cracking.

(b) The loss of bond which shows itself as a hollow sound when the surface is tapped may be located at the interface between the render and the concrete or between the coats of render.

(c) A diagram showing the extent of the debonding and the cracking should be drawn. This will give a clear picture of the extent of the defects and enable a practical specification for remedial work to be prepared.

(d) Some areas of rendering will have to be cut out, and this should preferably be done by a disc to help avoid damage to the adjacent render.

(e) The rendering selected for removal must be carefully cut out. The surface of the concrete must be prepared to receive the new render by exposure of the coarse aggregate and removal of all dust and grit.

(f) To help avoid errors in the assessment of 'suction' and to ensure good bond with the concrete, a bonding coat of cement/SBR slurry should be applied to the prepared concrete not more than 20 min before the application of the render.

(g) The mix for the rendering must be compatible with the concrete to which is to be applied. A very strong render should not be applied to a comparatively weak concrete.

(h) In any case, the mix for the first coat should not be richer than 1:4 by volume (about 1:4·7 by weight), cement to sand. It is beneficial to use an SBR emulsion in the mix as this will assist workability (and thus help to reduce the water content) and will improve bond and reduce permeability of the finished render.

(i) For subsequent coats, the mix should be weaker, say 1:4·5 by volume, and should be thinner. If the first coat is 12 mm, then the second could be 8 mm. Generally, it is better if the first coat is not less than about 10 mm. Some proprietary mixes for the finishing coat may be as thin as 4 mm.

(j) Each coat should be allowed to dry out and shrink as long as possible before the next coat is applied. For the first three days after application, precautions must be taken to prevent rapid drying out, and therefore protection from sun and wind must be provided. The period between successive coats of render will vary according to the weather, but a minimum of 3 days, and preferably 7, should be allowed.

(k) If, for any reason the thickness of the new render exceeds 25 mm, then a galvanised welded fabric mesh, securely anchored into the concrete, should be provided. An alternative, which the author favours, is to use austenitic stainless steel mesh, with the anchors in the concrete of the same material.

The recommendations given in (d)–(k) above refer to those areas where defective rendering has to be removed and replaced with new.

There may be occasions when the amount of debonding is small and the cracks are relatively inconspicuous. A decision on exactly how much work should be done in such cases can be difficult, and the client should be kept fully informed, particularly of the relative costs of renewal and repair.

If it is desired to effect some remedial work without actually cutting out any rendering, consideration can be given to the following:

Fine cracks can be 'grouted in' by application of a brush coat of cement and an SBR emulsion. After about 7 days from completion of this work, the whole area of rendering can be given one or more coats of a selected decorative sealing treatment.

For small isolated areas where loss of bond to the concrete has occurred, injection with an SBR emulsion may be successful in reducing or even eliminating the hollow sound when the area is tapped; it should also help to prevent the debonded area spreading.

As debonding and cracking are usually associated defects, it would be desirable to apply the decorative coating after all repairs have been completed.

BIBLIOGRAPHY

BRITISH STANDARDS INSTITUTION, BS 5385—*Code of Practice for wall tiling*, Part 2:1978, *External ceramic wall tiling and mosaics.*

BRITISH STANDARDS INSTITUTION, BS 5262—*Code of Practice for external rendered finishes.*

MONKS, W. and WARD, F., *External Rendering*, Cement & Concrete Association 47.102, London, 1982, p. 31.

SMITH, R. G., *Long term unrestrained expansion of test bricks*, Building Research Establishment, CP 16/73, May 1973, p. 5.

THOMAS, K., *Movement joints in brickwork*, Clay Products Technical Bureau, London, July 1966, Technical Note, 1 (10), p. 8.

Chapter 8

The Repair of Concrete Bridges

8.1 INTRODUCTION

While this chapter deals with defects and methods of repair, reference should be made to British Standard BS 5400—Steel, concrete and composite bridges, which between 1978 and 1984 was published in ten parts. Parts 1, 2, 4, 7, 8, 9 and 10 are relevant to concrete bridges. Anyone undertaking an investigation and specification for repair should be conversant with this Standard.

The basic principles which should be adopted for dealing with the deterioration of reinforced concrete bridges are essentially the same as for other reinforced concrete structures which require investigation, diagnosis and the preparation of a specification for the repair.

The Bridge Inspection Guide published by the Department of Transport in 1983 states that there are some 150 000 highway bridges in the United Kingdom, with a replacement value of about £15 000 000 000. It is believed that of these 150 000 about 50 000 are concrete.

The bridges which are considered in this chapter are road bridges; the publications listed in the Bibliography at the end of this chapter provide a great deal of essential information.

In addition to the Bridge Inspection Guide of the DoT, mentioned above, the following also deal with inspections:

(a) The Report by the Organization for Economic Co-operation & Development, 1976, on bridge inspection procedures in various countries.
(b) Department of Transport, Technical Memorandum BE4/77. Department of Transport, Technical Memorandum BE327.

Other relevant DoT publications include:

(c) Specification for Road & Bridge Works, 1976, with Supplement No. 1 and revisions.
(d) Notes for Guidance on Specification for Road and Bridge Works, 1976.
(e) Technical Memorandum (Bridges) BE27, Waterproofing and Surfacing of Bridge Decks.

The OECD Report and the DoT Technical Memorandum BE4/77 list and define four categories of inspection, namely:

Superficial Inspection
General Inspection
Principal Inspection
Special Inspection

The 1983 Bridge Inspection Guide deals mainly with General and Principal Inspections which are discussed collectively under the term 'Routine Inspections'.

General Inspections should be made at about two yearly intervals and the Principal Inspections about every 6 years. The Guide covers bridges constructed of materials other than concrete.

The defects which can and do develop in reinforced concrete structures generally have been described in some detail in Chapters 3, 4 and 5, together with the main reasons for their occurrence and recommendations for repair.

While accepting that basic principles for investigation and repair of reinforced concrete also apply to bridges, it must be kept in mind that bridge structures are subjected to quite different operating conditions than buildings and civil engineering structures. These differences must be reflected in the investigation, diagnosis and method of repair.

The fact that clear and authoritative recommendations exist for regular inspection of bridges is reasonable and helpful. No such recommendations exist for reinforced concrete buildings and many types of civil engineering structures.

This chapter will deal briefly with the following aspects of the deterioration of concrete highway bridges and will include notes on remedial work. The parts of a concrete bridge which are particularly vulnerable to deterioration include the following:

Joints—movement and sometimes construction joints. Waterproofing on bridge decks (if this fails the deck becomes very vulnerable).
Parts of the superstructure exposed to de-icing salt solution and spray.
Bridge bearings.

8.2 THE MAIN FACTORS CAUSING DETERIORATION

These may be summarised as follows:

(a) General and unavoidable wear and tear; increase in weight of traffic.
(b) Damage caused by accident/impact.
(c) Movement of the foundations.
(d) Substandard workmanship/materials; inadequate supervision during construction.
(e) Poor detailing, especially at bearings and joints, and poor arrangements for drainage.
(f) Inadequate waterproofing of deck, allowing ingress of water and de-icing salts containing chlorides.
(g) Sulphate attack on the substructure.
(h) Alkali–aggregate reaction.

While motorway pavements have for many years been constructed in air-entrained concrete it is often found that the superstructures of bridges are in non-air-entrained concrete. As a result damage has occurred to those parts of the superstructure subjected to splash from vehicles. This splash can cause physical deterioration due to the freezing of the water in the surface layers of the concrete, and also due to the penetration of chlorides (from de-icing salts) into the concrete down to the steel reinforcement. The provision of a carefully detailed membrane for the bridge deck should prevent penetration of water and chloride downwards into the deck. If there are defects in the membrane then serious corrosion can occur before it is detected because in addition to the membrane there is usually an asphalt wearing carpet. Careful attention to drainage is also needed.

Substandard workmanship is more common than the use of substandard materials, and usually takes the form of displaced reinforcement, resulting in inadequate cover, and lack of care in the compaction and curing of the concrete. Even the best quality concrete from the mixer will provide little long term protection to the reinforcement unless it is thoroughly compacted and properly cured.

In the US and Canada the use of de-icing salts (sodium and calcium chloride) has caused enormous damage to bridge structures. It has been reported that in the Syracuse area of New York State, some 10 000 tons of salt are applied to the highways during each winter. In the State of Michigan, normally de-icing operations are carried out 30–35 times each winter. Each storm may require 1–3 applications of de-icing salt. In the

UK, damage to concrete bridges due to de-icing chemicals is much less severe than in the US and Canada; nevertheless it is becoming more significant each year and damage varies from one part of the country to another.

Alkali–aggregate reaction is a comparatively new threat to the durability of concrete bridges in the UK. The problem was highlighted in a short article in the *New Civil Engineer* of 2 May 1985. The first cases of AAR occurred in the South West of England in the late 1970s, but since then examples have been found in the Midlands and further north. The *New Civil Engineer* stated that official estimates put the number of affected road bridges at over 350. On the other hand, the Cement & Concrete Association in October 1985, put the total figure of confirmed cases at about 60. Some general comments on AAR are given in Chapter 2, Section 2.2.6.

8.3 INVESTIGATIONS

These should proceed on the same lines as for a building structure which have been described in some detail in Chapter 3 and tests for chlorides should always be included. A half-cell survey is also essential as the top surface of the bridge deck is invariably covered with asphalt and frequently with a membrane as well. This means that serious corrosion of the top steel can occur without visible signs.

Cracks should be carefully scrutinised with the object of deciding on the type, and thus the likely cause of the cracking. Apart from cracks caused by overload or underdesign, the majority of cracks arise from drying shrinkage and thermal contraction. These two types of cracks have been discussed in some detail in Chapter 3, Section 3.2 to which readers are referred.

The existence or absence of bearings should be checked, and if present they should be carefully inspected to obtain information on the following:

(a) General condition.
(b) Conditions which may result in jamming, looseness, or limitation on movement.
(c) Condition of seating, bedding or other supports.

The condition of movement joints should receive detailed attention. These joints will be:

Full structural movement joints.
Contraction joints.
Sliding joints.

The joints should be carried through any finishings and must be properly detailed and sealed with a flexible sealant capable of accepting the anticipated maximum movement (opening and/or closing) of the joint. Full structural movement joints (often called expansion joints) in bridge decks have to operate under severe conditions of exposure, impact and wear. In reading the numerous papers on bridge design and maintenance, it is clear that there is a fairly wide divergence of views on the location and detailing of such joints.

In the past, finger joints, also known as comb joints, were used, but these have fallen into disfavour. Reports from the UK, US and Canada indicate that this type of joint introduces serious maintenance problems. The 'fingers' get clogged with hardened debris and de-icing salt which prevents the joint functioning properly. When the joints are kept open, salt laden water can pass through onto the bridge structure below.

Contraction joints are usually located in retaining walls and abutments. They should be properly sealed otherwise unsightly water staining will occur.

Sliding joints are really bearings, which may consist of one member sliding on another, or roller bearings or bearings which deform horizontally but are stiff enough to carry the vertical loads. Simple sliding bearings are usually polished steel plates some of which incorporate a low friction material such as polytetrafluoroethylene (PTFE).

All types of bearings need careful inspection. Major repairs are costly and difficult to carry out and propping and/or temporary closure of the bridge may be required.

Repairs to joints may involve the use of proprietary materials or systems. It is important to ensure that contractual responsibilities are clearly defined; suppliers should be contractually associated with the applicators.

8.4 REPAIRS

8.4.1 General Considerations
Repairs to bridges are much more difficult to carry out than repairs to buildings because every effort has to be made to avoid or reduce

disruption to traffic flow. This results in great pressure to open the bridge completely to traffic as quickly as possible, involving shift work, difficulties with supervision by both contractor and employer, and the use of materials which will reach the required strength in the shortest possible time.

What may be termed straightforward repairs to concrete should be carried out essentially as described in Chapters 4 and 5 and these may include the use of gunite. Special problems arise from the presence of chlorides in the concrete due to de-icing salts, defects in the waterproof membrane, and alkali–silica reaction (ASR). These are commented on below.

Chloride contamination of bridge decks has so far not occurred very often in the UK, but corrosion of reinforcement in bridge parapets and kerbs is not uncommon. In the US and Canada great damage has been caused to bridge decks and superstructures by chloride corrosion of rebars. For the past 20 years active research has been carried out in these two countries to try to find a practical and technically satisfactory solution to the problem.

This research examined two aspects of the problem, namely how to prevent the deterioration in the first place and how to do the repairs so that the bridge will be rehabilitated for an acceptable period. This book is only concerned with remedial work, and the techniques which give the most promise are (a) cathodic protection, (b) the removal of the concrete down to and around the rebars (the concentration of chloride decreases rapidly with increase in depth from the surface). The removal of chlorides by electrochemical means was tried in the US and Canada with mixed results, and does not appear to have gained much favour. Reports on various methods of repair have been published; one of the most interesting is by Manning and Ryell of the Ministry of Transportation & Communications, Ontario. In the US major work on the same problems has been carried out by Spellman and Stratfull of the California Dept. of Highways.

8.4.2 The Use of De-icing Chemicals

Parallel to the research into repair techniques, work has proceeded on finding an effective de-icing salt which is chloride free, is not aggressive to concrete, and which will not increase the alkali concentration of the concrete (for the significance of the latter, see comments in Section 8.4.8).

Urea $(CO(NH_2))$ has been used for some years as a de-icer on civil and military airfields, but opinions differ on the advisability of its use.

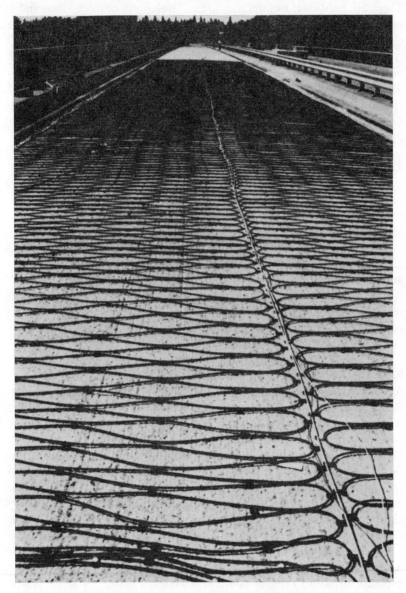

FIG. 8.1. Cathodic protection of a salt contaminated bridge deck by Ferex anodes and impressed current, prior to the laying of a concrete topping (courtesy: Raychem Ltd).

There are published reports that urea in water will corrode steel, and at certain concentrations may be more aggressive than sodium/calcium chloride (see the Bibliography at the end of this chapter).

Research in Norway (not yet published) has shown that a suitable neutral de-icing salt is a technical possibility and is undergoing trials. The author feels that when new compounds are being assessed, consideration should be given to the damage done to the bodywork of motor vehicles as well as to reinforcement in concrete, by the salts at present in use.

8.4.3 Cathodic Protection of Reinforcement

It is the author's opinion that the technique which holds the greatest promise of success in stopping chloride induced corrosion of reinforcement and providing a long term solution to the problem is cathodic protection.

Two systems of cathodic protection have been briefly described in Chapter 4, Section 4.6. Figure 8.1 shows one such system using Ferex anodes laid out on a bridge deck in the US. The anodes will be covered with a suitable concrete topping as a running surface. These techniques will be modified and improved in the light of site experience. The main problem is to ensure that the steel is fully protected with the use of the minimum amount of impressed current.

8.4.4 The Removal of Chloride-contaminated Concrete

An interesting report on the repair of a chloride-contaminated bridge deck using the orthodox method of concrete removal and replacement, was given in a paper by Cavalier and Vassie in the *Proc. ICE*, August 1981 (complete reference is given in the Bibliography at the end of this Chapter). The paper gives a considerable amount of useful information on how the investigation was conducted. This included coring to check the depth of chloride penetration and a half-cell survey to locate and map corrosion intensity. There is also a section on acceptable levels of chloride in concrete with special reference to 'combined' and 'free' chloride ions.

On completion of the repair, resistance probes were left in position to monitor future corrosion. Prior to placing new concrete the rebars and exposed surface of existing concrete were given a slurry coat of cement and sand.

8.4.5 Defects in the Bridge Deck Waterproof Membrane

Prior to the end of the World War II very few bridge decks in the UK were provided with a waterproof membrane. The provision of such a membrane only became mandatory in about 1965.

Waterproof membranes on concrete bridge decks are required because the surfacing, such as mastic asphalt, dense bituminous macadam, etc., is permeable. The concrete deck must be protected against ingress of water, particularly solutions of de-icing salts such as sodium and calcium chloride. Mention has been made in Section 8.4.2 that urea may also be aggressive to steel reinforcement.

The membrane must possess long term durability; in theory it should last the life time of the bridge structure. It is essential that the membrane should extend over the whole of the bridge deck including areas covered by footways, verges, and central reservation. Special care is required to ensure that the membrane is properly detailed and sealed at the whole of the perimeter, and around gulleys and around protrusions from the concrete deck, and at all movement joints.

The Department of Transport require that the membrane should either comply in full with clause 2643 of the Specification for Road and Bridge Works, or the details set out in Technical Memorandum No. BE27. Membranes which are less than 12 mm thick must be protected immediately after laying; protection must always be provided where renewal of the wearing course may necessitate burning off. Technical Memorandum BE27 lists nine 'acceptable' waterproofing materials, with a note that the list will be extended as other materials are approved. Waterproof membranes are proprietary materials/systems, and the DoT normally require that the material/system possesses an Agrément Certificate (issued by the British Board of Agrément). Firms supplying the materials/systems provide complete details of the method of protection and method of application. Figure 8.2 shows a heavy duty rubber-bitumen sheet membrane being laid on a concrete bridge deck.

In the UK the absence of membranes, or defects in the membrane caused by damage or poor detailing, has resulted in corrosion of reinforcement in some bridge decks, but the scale of damage is small compared with what has happened in the US and Canada and this has been commented on earlier in this chapter.

8.4.6 The Use of Polymer Resins for Remedial Work

There are many ways in which polymer resins can be effectively used in remedial work on bridges, but this does not mean that they are the answer to every problem. The use of a slurry composed of Portland cement and styrene–butadiene emulsion, mix proportions of two parts cement to one part emulsion (or one to one), can materially improve

FIG. 8.2. Heavy duty Bituthene rubber/bitumen sheet membrane being laid on a concrete bridge deck (courtesy: Servicised Ltd).

bond between existing concrete and new concrete and mortar. When applied to cleaned rebars it provides an intense alkaline environment which passivates the steel. It is the author's opinion that it is in most cases better to use such a coating rather than an inert barrier of epoxy resin. This is particularly important when sections of a reinforced concrete unit are being repaired; the application of the cement modified slurry restores the original passivation and is compatible with the condition of the rebars which remain embedded in the existing concrete. The provision of an inert coating on isolated sections of rebars can result in significant variations in potential on the surface of the steel. It is a different matter if the whole of a rebar can be completely coated, and mention has been made in Chapter 1, Section 1.4 of epoxy powder coated rebars.

Another use for polymers is as admixtures in concrete and mortar. The principal polymers used are styrene–butadiene rubber (SBR) and acrylic emulsions. The author's experience has been mainly confined to SBRs and he has found the addition of 10 litres of SBR to each 50 kg cement in the mix to be effective in reducing shrinkage and reducing permeability. The reduction in drying shrinkage arises from the fact that the SBR acts as a workability aid (lower water/cement ratio with constant workability).

The decision to use epoxy mortars for repair should be taken with care, particularly if the concrete to which they will be bonded is of rather low strength as the strength of the epoxy mortar may be very high. A further point is that the coefficient of thermal movement of the polymer mortar is appreciably higher than that of concrete (20–30×10^{-6} for an epoxy mortar and 7–13×10^{-6} for a cement mortar or concrete).

A third use for polymers is in crack injection; the principal resins being epoxy and polyurethanes. Some information on crack injection of concrete has been given in Chapter 5, Section 5.2, to which readers should refer. Crack injection can only, at the best, restore the strength of the member to what it was originally. Where the cracks arise from drying shrinkage, plastic settlement cracking or thermal contraction (see Chapter 3, Section 3.2), then injection, properly carried out, should prevent water penetration into the crack. With specially formulated resin and experienced application cracks down to a width of 0·015 mm can be successfully injected.

An effort should be made to prove the effectiveness of the injection by cutting cores on the line of some of the cracks and checking the actual penetration of the resin.

8.4.7 The Use of Steel Plates to Strengthen Bridge Beams

It appears that the technique was first used in France and S. Africa in the late 1960s and early 1970s. Since then its use in Switzerland and Japan has been recorded. In the UK, the Transport and Road Research Laboratory has carried out research and testing starting in about 1976. In 1979, K. D. Raithby of TRRL, Crowthorne, submitted a paper to the American Concrete Institute Symposium on Strength Evaluation of Existing Structures at Washington DC. This described the original research work at TRRL and the strengthening of four bridges at an inter-change where a motorway (M5) crosses a trunk road (A456). The report by K. D. Raithby which is an expanded version of his paper for the ACI Symposium, is included in the Bibliography at the end of this chapter. The main fact about this technique is that success relies on first class bond between the soffit of the concrete member being strength-ened and the steel plates. The object is to ensure that the concrete and steel plates act compositely.

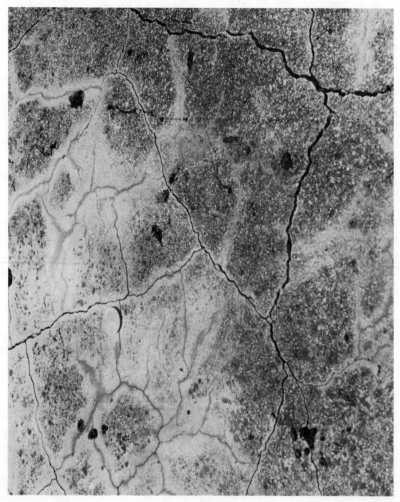

FIG. 8.3. Map cracking typical of alkali–silica reaction. (Crown copyright. Reproduced from Building Research Establishment Digest 258 by permission of the Controller of HMSO.)

8.4.8 Alkali–Silica Reaction

This type of deterioration of concrete in the UK has received a great deal of publicity in recent years. In Section 8.2 it was mentioned that there was reason to believe that some 350 road bridges may be at risk, and that steps were being taken to carry out detailed investigations.

At the present time there is no known method of stopping the reaction between the cement and the aggregates; all that can be done is to remove or reduce those conditions which are likely to aggravate the situation; see Fig. 8.3 which shows cracking typical of alkali–silica reaction. The alkalinity of the cement paste cannot be reduced in hardened concrete, neither can the special characteristics of the siliceous aggregates be changed or modified.

It is known that warmth and moisture increase the rate of the reaction and hence the rate of deterioration. It therefore follows that steps to reduce temperature rise in the concrete members will help, even if only a little. Depending on the colour of the existing concrete, the application of a white or light coloured coating would help to control temperature rise due to solar gain.

Probably the single most effective measure is to ensure the best possible drainage of surface and rain water so that there is no ponding, and run-off is as rapid and complete as possible. The provision of a water-resistant coating can also be useful in reducing water penetration into the concrete, but it should be applied after a spell of dry weather or after artificial drying out of the surface layers. Also, the coating should possess high vapour permeance, but this is difficult to combine with high water resistance and long term durability.

It must be emphasised that the two measures mentioned above (reduction in solar gain and reduction in water penetration) will only mitigate the effect of ASR by slowing down the rate of attack, and they will not stop it.

The most drastic measure, which could only be adopted in special cases when site conditions are suitable, is to encase completely the affected concrete members in thick, heavily reinforced concrete or gunite/shotcrete. A structural assessment would have to be made to assess the effect of the increase in dead load, and for the design of the new reinforcement.

Authoritative publications on this complex subject stress that it is the total alkalis in the concrete which are important although it is known that these originate almost entirely in the cement. However, if there are external sources of solutions of alkalis such as sodium salts, e.g. sodium chloride, and these penetrate into the concrete, ASR may be aggravated or even triggered. This is being investigated at the time this book was written.

BIBLIOGRAPHY

ALDEEN, S., Comb joint trouble closes flyover, *New Civil Engineer*, London, 21 July, 1977.

AMERICAN CONCRETE INSTITUTE, Bridge decks, repair and maintenance, five papers, *ACI Journal*, Dec. 1975.

AMERICAN CONCRETE INSTITUTE, *Guide for repair of concrete bridge superstructures*, ACI Committee 546, ref. 546.1R-180, 1980, p. 20.

AMERICAN CONCRETE INSTITUTE, *Joint sealing and bearing systems for concrete structures*, ref. SP-70, 2 vols, 1981.

AMERICAN SOCIETY FOR TESTING AND MATERIALS, *Standard test method for half-cell potentials of reinforcing steel in concrete*, ANSI/ASTM C876–80, pp. 548–54.

ANON, Report pinpoints crack attack. *New Civil Engineer*, June 12, 1986, pp. 24–5.

BRITISH STANDARDS INSTITUTION, *The structural use of concrete*, BS 8110, Parts 1 and 2:1985 (replaces CP 110).

BRITISH STANDARDS INSTITUTION, *Recommendations for non-destructive methods of testing concrete*, BS 4408.

BRITISH STANDARDS INSTITUTION, *Guide to the assessment of concrete strength in existing structures*, BS 6089.

BRITISH STANDARDS INSTITUTION, *Methods of testing concrete*, BS 1881, Parts 1–6 and 101 to 122.

BRITISH STANDARDS INSTITUTION, *Steel, concrete and composite bridges*, BS 5400, Parts 1 to 10.

BUILDING RESEARCH ESTABLISHMENT, *Determination of chloride and cement content in hardened Portland cement concrete*; Information Sheet IS 13/77, July 1977, p. 4.

BUILDING RESEARCH ESTABLISHMENT, *Alkali–aggregate reactions in concrete*, Digest No. 258, Feb. 1982, p. 8.

CAMPBELL-ALLEN, D. AND LAU, B., *Cracks in concrete bridge decks*, Commissioner for Main Roads, New South Wales, Australia.

CAVALIER, P. G. AND VASSIE, P. R. W., Investigation and repair of reinforcement corrosion in a bridge deck. *Proc, Inst. Civ. Engrs*, Part 1, August 1981, 461–80.

CEMENT & CONCRETE ASSOCIATION, *The Hawkins Working Party Report on alkali–aggregate reaction; minimising the risk of alkali–silica reaction— Guidance Notes*. Published by the Association, no. 97. 304, Sept. 1983, p. 8.

CONCRETE SOCIETY, *Concrete core testing for strength*, Technical Report No. 11, 1976, p. 44.

CONCRETE SOCIETY, *Concrete bridges, investigation, maintenance and repair*, Symposium, London 25 Sept. 1985, eight papers, published by the Concrete Society, London.

DEPARTMENT OF TRANSPORT, *Specification for Road and Bridge Works*, 1976, with Supplement No. 1 and Revisions, HMSO, London, pp. 194 and 66.

DEPARTMENT OF TRANSPORT, *Bridge inspection guide*, HMSO, 1983, p. 52.

DEPARTMENT OF TRANSPORT, *Notes for Guidance on Specification for Road and Bridge Works*, HMSO, London, p. 128.

DEPARTMENT OF TRANSPORT, Technical Memorandum (Bridges) BE27, *Waterproofing and Surfacing of Bridge Decks*, DoT, London.

FEDERAL HIGHWAY ADMINISTRATION, *Styrene–butadiene latex modifiers for bridge deck overlay concrete*, Report FHWA-RD-78-35; reproduced by National Technical Information Service, April 1978, p. 105.

FROMM, H. J. AND WILSON, G. P., *Cathodic protection of bridge decks—a study of three Ontario bridges*, Ministry of Transportation and Communications, Ontario, 1980, p. 48.

GEE KIN CHOU, *Rebar corrosion and cathodic protection—an introduction*, Raychem Corp., California, p. 11.

HOBBS, D. W., The expansion of concrete due to alkali–silica reaction, *Structural Engineer*, Jan. 1984.

JERMAN, J. J., Cost control of alternative ice control methods, *Public Works*, US, Feb. 1976.

MCANOY, R., BROOMFIELD, J. P. AND DAS, C. S., *Cathodic protection—a long term solution to chloride induced corrosion?* Paper at International Conference, Structural Faults '85, Institution of Civil Engineers, London, April/May 1985, p. 7.

MACDONALD, M. D., *Waterproofing of concrete bridge decks*, TRRL Lab. Report 636, 1974, p. 32.

MANNING, D. G. AND RYELL, J., *Durable bridge decks*, Ministry of Transportation and Communications, Ontario, 1976, p. 67.

MANNING, D. G. AND RYELL, J., *A strategy for the rehabilitation of concrete bridge decks*, Ministry of Transportation and Communications, Ontario, 1980, p. 30.

RAITHBY, K. D., *External strengthening of concrete bridge beams with bonded steel plates*, Transport and Road Research Laboratory, Crowthorne, Supplementary Report 612, 1980, p. 18.

SCHRADER, E. K. AND MUNCH, A. V., Deck slab repaired by fibrous concrete overlay, *American Society of Civil Engineers, Journal of the Construction Division*, Mar. 1976, pp. 179–96.

SLATER, J. E., LANKARD, D. R. AND MORELAND, P. J., Prevention of rebar corrosion by electro-chemical removal of chlorides from concrete bridge decks. *Corrosion*, 1976, Paper No. 20, 16.

SOMERVILLE, G., *Engineering Aspects of Alkali–Silica Reaction*. Cement & Concrete Association, Interim Technical Note 8, Oct. 1985, p. 7.

SPELLMAN, D. L. and STRATFULL, R. F., *Laboratory corrosion test of steel in concrete*, California Department of Public Works, Division of Highways, Interim Report M & R No. 635116-3, September 1968, p. 39.

STRATFULL, R. F., Half-cell potentials and corrosion of steel in concrete, *US Highway Research Record*, 1973, p. 12–21.

TRANSPORTATION RESEARCH BOARD, *Bridge deck joint-sealing systems*, National Co-operative Highway Research Program Report 204, June 1979, p. 31.

VASSIE, P. R. W., *Evaluation techniques for investigating the corrosion of steel in concrete*, Transport and Road Research Laboratory, TRRL Supplementary Report 397, 1978, p. 22.

WATSON, S. C., *Zero maintenance expansion joints and bearings, a design goal for the future*, FIP 8th Congress, London, 1978. Published in New York by Watson Bowman Associates, p. 56.

Chapter 9

Repairs to Concrete Water-Retaining and Water-Excluding Structures

9.1 INTRODUCTION

The usual reason for repairing a water-retaining or water-excluding structure is to remedy leakage. To avoid unnecessary repetition, these two types of structure will be referred to throughout this chapter as water-retaining structures and basements respectively.

The leakage may be:

(a) outward from the structure, e.g. from reservoirs, swimming pools, sewage tanks etc.;
(b) outward when the structure is full and inward when it is only partially full or empty, e.g. reservoirs, etc.;
(c) inward from the surrounding ground, as in the case of a basement.

Associated with any form of leakage, particularly when the structure has been in use for some years, is likely to be the corrosion of the reinforcement and spalling and cracking of the concrete.

It must be realised that in practice, no concrete structure will be what is known as 'bottle tight', unless it is lined with a waterproof membrane. When the structure is properly designed and constructed the amount of leakage should be very small; the amount that can be tolerated will depend on the circumstances of each case. Loss of water from water-containing structures is closely connected with a leakage test, which is normally applied to all new structures. This test is discussed in some detail later in this chapter.

In the case of basements there can be no prescribed water test as such, but basements are expected to be dry. Although in theory, a basement in reinforced concrete can be constructed so as to be completely watertight without resort to tanking, the author's experience is that this desirable aim is unlikely to be achieved in practice. Those responsible for the design of a basement should give very careful consideration to the standard of watertightness required; the design and specification should then be prepared accordingly.

A fundamental principle of repair is to seal the leaks on the water face. Unfortunately this is not always possible, and specialised firms have developed techniques for sealing leaks on the 'wrong' side, that is against the flow of water. However, due to the basic difficulties in executing such repairs, guarantees of complete success are seldom offered; if they are, they should be read with great care. When it is proposed to carry out waterproofing work to the inside face of basements, consideration should be given to the possible use of pressure grouting to supplement the surface sealing. The object of the grouting is to help prevent the water penetrating into the body of the wall.

In the case of old reservoirs, the argument is sometimes advanced that the cost of the water lost by leakage is small compared with the cost of the necessary repairs. This may well be so, but consideration should also be given to the possible effect on the foundations and stability of the structure, by the continuous flow of several thousand gallons per day through the walls and floor.

9.2 THE WATER TEST ON WATER-RETAINING STRUCTURES

Almost all new water-retaining structures, such as reservoirs, swimming pools, sewage tanks, etc., are required to pass a water test. Repairs are undertaken when the structure fails to pass the test. In the case of existing structures, when leakage is suspected, it is also usual to carry out a water test. If it fails the test, then investigations are put in hand to determine the location of the leaks, assess why they have occurred, and arrange for their repair. After the repairs are completed, a further test is carried out. Because of this close connection between water test and repairs, the author feels that some discussion on the water test would be useful.

The Code of Practice, BS 5337, states: 'The engineer should specify the permissible drop in the surface level taking into account the losses due to absorption and evaporation. For many purposes the structure may be deemed to be watertight, if the total drop in surface level does not exceed 10 mm in 7 days ... For open reservoirs tested in a similar manner an additional allowance should be made for loss due to evaporation.' The Code recommends that roofs be tested for watertightness when the structure holds potable water. Domed roofs should be provided with a waterproof membrane.

The Code recommends that a flat concrete roof should be tested by lagooning the roof slab to a minimum depth of 75 mm for a period of 3 days. The roof slab can then be regarded as satisfactory if no damp patches occur on the soffit during this period.

It would appear from the wording of the Code that the intention is that the concrete roof slab would be watertight in its own right and that except for domed roofs a membrane is not required. The author considers that the conditions of the test are very severe as in practice it is virtually impossible to construct a flat roof slab so that no damp patches appear unless a membrane is used.

The author is in agreement with the principle of testing the flat roofs of structures holding drinking water before a waterproof membrane is laid on the roof slab. However, it is prudent to provide a membrane, and the object of the water test would be to detect any leaks which should then be repaired prior to the application of the membrane. Damp patches would under these circumstances not be considered as 'leaks'.

Regarding damp patches, all engineers experienced in reservoir construction know that there is usually condensation on the soffit of the slab. The presence of these patches could cause a dispute to arise. Therefore the slab should be checked for condensation before the roof is ponded; even then, the presence of the water on the roof may result in a drop in temperature of the slab and if it then fell below the dew point, condensation would occur.

While the test is undoubtedly useful, the inherent difficulties mentioned above should be recognised and dealt with in a practical way.

For elevated structures, such as water towers and swimming pools above ground level, where the outside of the walls and the underside of the floor can be inspected, the usual test requirement of maximum drop in level over seven days can be substituted by a clause requiring that there should be no visible sign of leakage.

In considering the Code recommendations, it should be noted that they are not clear. Having said that the engineer should take into account absorption and evaporation, it limits the *total* drop in surface level. In the next paragraph it says that in the case of open reservoirs an additional allowance should be made for evaporation. In the opinion of the author it is better to avoid disputes on site and arrange in the test requirements for the practical measurement of evaporation from both open and closed structures.

Metropolitan Water Board (now Thames Water Authority) engineers have found that there can be appreciable evaporation in roofed service reservoirs.

A simple and effective method of measuring evaporation is to fill a steel/plastic drum with water to within about 50 mm–75 mm of the top, and to anchor this in the water of the reservoir during the period of the test. It can be assumed that the drum is completely watertight and therefore any drop in level in the drum will be due to evaporation.

There are a number of theoretical arguments which can be used to show that the evaporation from the drum is different to that from the water surface of the reservoir. No doubt there is some difference, but this is likely to be smaller than the large discrepancy which would result from the use of a formula and the insertion of factors which have to be evaluated by guesswork.

When a water-retaining structure is being filled for the first time (this is usually for the water test), the water level should rise slowly. The author's opinion is that a rate of 0·75 m/24 h should not be exceeded. The majority of structures of this type do not contain more than a 7 m depth of water so that such a structure can be filled in 10 days.

As soon as the structure is full to overflow level, which is usually above operational top water level, the level should be accurately recorded. If a high degree of accuracy is thought necessary, then a hook gauge can be used; otherwise the level can be recorded on the wall or other accessible part of the structure.

The level should be read daily at the same hour; the drop in level should reduce each day so that at the end of the initial 'soakage' period of 7 days, the recorded 24 h drop should not exceed about 2 mm, which is difficult to measure anyway. In certain cases, particularly when the structure has been under construction for a long period, in dry weather, the initial period may have to be extended to 10 or 14 days before relative stability is achieved.

The 7-day test period should then be started by recording the water level in the structure and in the barrel used to record the loss due to evaporation. It is advisable to check both levels each day at the same hour.

If the drop in level over the 7-day test period does not exceed the figure in the specification and there are no signs of seepage on any of the exposed surfaces, the structure can be assumed to have passed the test.

The question is what should this 'figure in the specification' be. As previously stated, the recommendations in the Code are ambiguous. Many experienced water engineers consider that the loss of water, excluding evaporation, should be appreciably less than 10 mm. It is better to decide on a figure excluding evaporation and insert this clearly in the specification. It is useful to realise the quantity of water lost by a drop in level of say 10 mm; this amounts to $1 \, m^3/100 \, m^2$ of water surface. This is an appreciable quantity of water; seepage through fine hair cracks and slightly porous concrete would certainly not result in this amount of loss over a period of 7 days. It is therefore important to include a requirement in the specification that there shall be no visible signs of leakage on any exposed surface of the structure. This will make it clear that all such defects must be put right under the contract without extra payment.

9.3 TRACING LEAKS

The next problem is how to trace and locate leaks when the drop in water level exceeds that allowed in the specification. This should present no difficulty with elevated structures, but for those on or in the ground where the underside of the floor cannot be inspected, and sometimes the lower part of the walls as well, considerable difficulty can arise.

The author has recommended in his book *Swimming Pools*, that it is worthwhile to spend a little more money than is usual on the under-drainage of the floor. This will enable the underfloor system of drains to be laid in such a way that the floor area is divided into a number of separate units as shown in Fig. 9.1. If the external walls are embanked, it is usual practice for a perimeter drain to be laid just below kicker level to drain the embankments. If this perimeter drain is kept separate from the underfloor drainage it will enable leakage through the wall to be seen separately to any flow arising from leakage through the floor. By dividing the underfloor drainage into units, each controlled by a manhole, flow in the system can be monitored, and the search narrowed to a group of about six floor bays.

As previously stated, the most likely places for leakage in a floor are through the joints and these should be carefully examined. Unfortunately, the author's experience is that there can be considerable seepage through joints without any visible sign of defects on the surface.

All this sounds very depressing, and to remedy leaks in floors can be

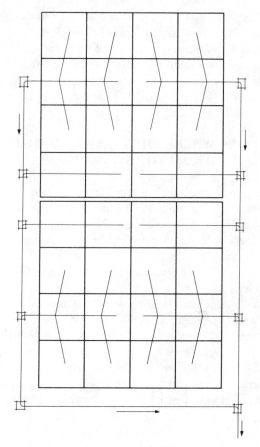

FIG. 9.1. Suggested layout of underdrainage of reservoir floor to assist in location of leakage.

very time consuming and consequently expensive. The Hydrology and Coastal Sedimentation Group of the Atomic Energy Research Station at Harwell, developed techniques in the 1960s for tracing leaks by means of short-lived radioactive tracers. The Group carries out the work itself and is responsible for the selection of the tracers used in each case. The technique is expensive, but the standard routine of trial and error may cost even more if accurate costing procedures are adopted. The tracer technique is particularly useful where there is no underdrainage, or the drainage system has no access points.

Where underdrainage, as suggested above and shown in Fig. 9.1, has been provided, dyes, concentrated salt solution, etc., can be used to supplement visual inspections at manholes. However, it must be remembered that dyes will stain the surface of the concrete as well as most other materials. This is important when the structure has already been given a decorative finish, for example a swimming pool.

9.4 REPAIRS WHERE THE LEAKAGE IS OUTWARDS FROM THE STRUCTURE

Water-retaining structures such as reservoirs, swimming pools, sewage tanks, etc., are normally designed and constructed in accordance with the British Standard Code of Practice, BS 5337.

Leakage generally arises from three main causes: cracks; porous and/or honeycombed concrete, defective joints. In most cases this type of leakage is repaired with the structure empty, and the repair is effected on the inside (water face).

However, with structures holding industrial liquids, the emptying of the structure and the execution of the repair on the inside would cause great dislocation to the operation of the plant. In such cases the owner wishes the repair to be effected with the structure in operation. This means that the repair has to be effected against the pressure of the contained liquid. The author's experience is that it is seldom that such a repair results in achievement of even reasonable watertightness, unless, after all individual areas of seepage have been effectively sealed, the whole of the outside is given a coat of structural gunite of adequate thickness.

If the work is carried out by an experienced contractor (who is unlikely to be the lowest tenderer if there is competitive tendering!), then considerable improvement can be reasonably expected, but the author doubts whether such a contractor would give a guarantee for complete elimination of seepage.

From time to time materials are offered on the market which are claimed to seal leaks against the pressure of the contained liquid by the process of the growth of crystals into the concrete. It is claimed that the crystal growth forces back the water in the pore structure of the concrete and the growth takes place after the proprietary mortar has been applied to the outside of the structure and has hardened. The author has not seen any evidence to support the claim of crystal growth inwards from an externally applied mortar or slurry. In hydrated cement paste in hardened concrete there are two distinct classes of pores, namely capillary pores and gel pores. The size of capillary pores is about $1.3\,\mu m$ (0·0013 mm) (a human hair is about 0·1 mm). Gel pores are much smaller, being of the order of 15–20 Angstroms (one Angstrom is 10^{-7} mm). In reasonable quality concrete capillary pores become discontinuous after some months. It seems to the author that the chances of crystal growth displacing the water in capillary pores is about zero.

9.4.1 Repair of Cracks

It has been stated in Chapter 5 that to repair a crack successfully it is first necessary to know why the crack has formed, and then to decide whether further movement is likely to occur which would cause the crack to reopen if it were repaired with a rigid material.

In new structures, cracking is usually more prevalent in the walls than in the floor and is generally caused by thermal contraction stresses in the early life of the concrete. This type of cracking has been discussed in Chapter 3, and a typical thermal crack in a cantilever slab is shown in Fig. 3.4. Figure 9.2 shows such a crack in a reservoir wall.

FIG. 9.2. Typical thermal contraction crack in reservoir wall.

For roofed structures which are either in the ground or are embanked, the chances of significant movement, once the tank is in operation, are quite small, unless foundation movement occurs. Therefore cracks in this type of structure can be safely sealed with epoxide resins and similar materials which are rigid. It is advisable for the sealing to be carried out as late in the construction process as possible. When the structure is filled with water the cracks will tend to close due to moisture movement as the concrete becomes progressively wetter. In this way, the sealant in the crack, whether it is inserted or injected, will be put into compression. Unless the structure remains empty for a considerable time and the concrete dries out, the cracks should remain closed.

These thermal contraction cracks penetrate right through the wall or floor and therefore constitute a plane of weakness. If they are very fine and have practically sealed themselves while under test by the deposition of calcium carbonate (sometimes known as autogenous healing), they should be repaired as described below. While epoxide resins are normally rigid when cured, specially formulated resins can now be obtained which possess some degree of flexibility. Polyurethanes can be formulated so as to be very flexible. However, for drinking water reservoirs, the sealant should be non-toxic, non-tainting and should not support bacterial growth. Bitumen compounds, if used in large quantities, can, under some circumstances, impart a phenol taste to the water.

A decision must be taken as to whether the cracks will be repaired by crack injection methods, as described in Chapter 5, or by a crack filling and sealing technique. A brief description of the latter is as follows:

(a) Carefully tap down the line of the crack with a chisel so as to remove any weak edges caused by honeycombing behind. It should be noted that this is not the same as cutting out the crack. For cracks of this type, the author does not consider that cutting out is either necessary or desirable.

(b) Remove the laitance on the surface of the concrete for a distance of 300 mm on each side of the crack. This can be done by power operated wire brushes, light grit blasting, high velocity water jets, or light bush hammering. All grit and dust must be removed from the prepared surface.

(c) Brush into the crack and onto the prepared area, a minimum of three coats of low viscosity epoxide resin. The resin must be formulated to bond to damp concrete.

Figure 9.3 shows the above work in diagram form.

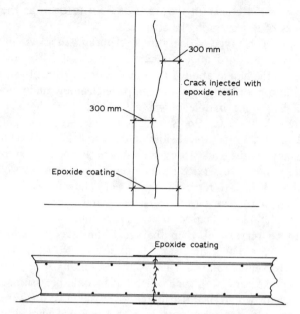

FIG. 9.3. Method of sealing crack in r.c. wall or suspended slab.

The author is not in favour of sealing cracks by means of a strip of preformed material such as polyethylene, butyl rubber or polyiso-butylene, which is fixed to the concrete by an adhesive. The reason for this is that the adhesive is liable to fail within a year or so because both edges of the strip are exposed to continuous immersion.

Questions are sometimes raised on the subject of 'autogenous healing' of cracks in concrete. The author has not seen an authoritative definition of this term, but it is usually taken to mean that the crack seals itself. This can only occur with very fine cracks, probably less than 0·10 mm wide, provided there is no further movement at the crack. The author's experience is that this 'healing' either occurs within the first week or so or the crack does not really seal itself.

There are cases, however, where there is reason to believe that the cracks may open at a later date during the anticipated operating cycle of the structure. This can occur when there is likely to be an appreciable drop in temperature in a structure where provision has not been made for this in the form of movement or partial movement joints or additional reinforcement. The thermal contraction stresses which develop may cause repaired cracks in the walls or floor to open as they are of course planes of weakness in the structure.

An example of this type of operating cycle occurs in a swimming pool. The water is maintained at a temperature of about 27°C, and the air temperature in the hall at about 29°C, for a period of three to six or more years. Then the pool is emptied for complete cleaning, maintenance and repair, usually in the middle of winter. This can result in a drop in temperature of 20–25°C. Pools which are elevated are particularly vulnerable to this, as the area below and around the pool shell is often used for plant rooms and stores, etc. When the pool is emptied, the opportunity is taken to service the plant and equipment as well.

The problem is how to repair cracks so as to allow a certain amount of flexibility and at the same time to ensure watertightness, and this can present considerable difficulty. If the movement across the crack is expected to be very small, then the use of an epoxide resin specially formulated to provide a degree of flexibility is probably as good as anything at present available. This would be suitable when there are a number of cracks parallel to each other and it could be assumed that the total movement would be distributed over say two or three cracks instead of concentrated across one. In such a case, i.e. where it is decided to use a flexible epoxide resin, the crack should be cut out to a width of 20 mm and to a depth of 20 mm; this groove should then be carefully filled in as directed by the supplier of the resin.

Crack injection as described in Chapter 5, using a resin of low viscosity and low 'E', value may also provide an acceptable solution. It must be remembered that the sealing of cracks in the structure itself with a semiflexible material, does not in any way solve the problem of possible cracking in rigid materials, such as tiling or protective chemically resistant rendering, which may be applied later.

When it is considered that the possible movement across the crack will be greater than can be accommodated by the flexible epoxide resin, then another method should be adopted. The detailing of such repairs requires careful thought and the following is one solution, but this should not be considered as necessarily applicable to all cases.

(a) A channel should be cut in the concrete for the full height of the wall, so that the crack is within the cut-out section as shown in Fig. 9.4. The depth of this channel need not exceed 5 mm. The sides of the channel should be cut with a saw, and the surface of the channel should be smoothed with a carborundum wheel.

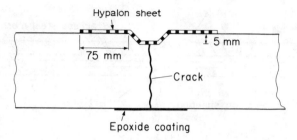

FIG. 9.4. Suggestion for sealing crack in wall.

Following this preparatory work all grit and dust must be removed.

(b) A low viscosity epoxide resin should be injected into the crack. The resin should bond to damp concrete and possess some degree of flexibility. Special care is needed for this type of injection. If the anticipated movement across the crack is likely to exceed the safe extension of the resin it is better not to use resin injection at all. The reason for this is that the resin may bond so strongly to the concrete that when tension develops across the crack, the concrete may fracture nearby, usually parallel to the original crack.

(c) Line the channel with Hypalon, or chlorinated polyethylene, and carry this along the face of the concrete to a minimum distance of 75 mm each side of the channel, all as shown in Fig. 9.4. The sheet should not be less than 1 mm thick and should be bonded to the concrete with an adhesive which is not moisture-sensitive, over the 75 mm distance.

If the walls are to be rendered or given some other applied finish such as ceramic tiles, then the detail will be somewhat different, and a suggestion for this is shown in Fig. 9.5. It should be appreciated that the crack in the wall will not be directly below the joint in the rendering and tiling and therefore complete absence of subsequent cracking in the rendering, etc., cannot be guaranteed. Figures 9.4 and 9.5 show resin coating across the crack on the outside face of the wall. Obviously this can only be applied if the outer face is accessible. The author is in favour of the use of a rigid (hard) resin for this external coating because if this fractures, it would indicate that movement across the crack was taking place and alert supervisory staff.

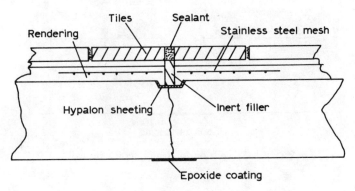

FIG. 9.5. Suggested method for sealing crack in tiled r.c. wall.

9.4.2 Repair of Defective Joints

The author's experience is that defective joints are the principal cause of leakage in water-retaining and water-excluding structures. Joints of any kind in this type of structure can be termed a necessary evil. It is impossible to build a liquid-retaining structure without joints; in addition, the joints must be correctly located and correctly detailed. It is unfortunate, to say the least, that some designing authorities leave matters relating to joints to the contractor and then blame him when leaks occur. This attitude is certainly contrary to the intention of the relevant clauses in Code of Practice BS 5337. The Code makes it clear that the location and detailing of joints is an essential part of the design and is the responsibility of the engineer. Joints detailed and designed to allow thermal movement to take place must be provided unless the engineer is satisfied that other satisfactory provision has been made to control thermal contraction cracking. There are numerous papers on this important subject and a reader who wishes information should refer to the Bibliography at the end of this chapter.

It is normal practice, although some engineers do not agree with this, to provide water bars in joints. In walls and suspended slabs, these are usually the centrally located dumb-bell type. In addition, the water side of the joint is usually provided with a sealing groove which is filled with a sealant. However, for what are known as monolithic (construction or day-work) joints, some engineers rely entirely on the bond between the old and new concrete for watertightness.

Figures 9.6 and 9.7 show defective joints in a reservoir before and during repair.

FIG. 9.6. Deteriorated sealant in reservoir wall (courtesy: Colebrand Ltd and Thames Water Authority).

In spite of these rather elaborate and expensive precautions, joints often leak, In the case of walls and suspended slabs, the location of the leak is clearly visible, but with floor slabs supported on the ground, the tracing of the leak is very difficult indeed. Most structures of this type are underdrained, and if the layout of the drainage system has been designed with this problem in mind (i.e. each section of the underdrainage deals with a particular area of floor), then the locating of the leak is much facilitated. Even so, there is likely to be an appreciable area of floor which is under suspicion. Once it is established that the leakage is

FIG. 9.7. Joints in Fig. 9.6 after resealing with 'Neoferma' (EPDM) gaskets (courtesy: Colebrand Ltd and Thames Water Authority).

through the floor, then in most cases it is fairly certain that it is the joints which are responsible. The flow through the underdrains must be carefully monitored as each length of joint is ponded. This is a time-consuming and difficult job and requires patience and care. Sometimes reference to the daily reports from the R.E. or Clerk of Works will be useful in revealing some forgotten hitch in the work which may have resulted in undercompaction of concrete around a water bar, or displacement of a water bar during concreting. It is not possible in a book to discuss all the possible trouble which can occur at a joint and so

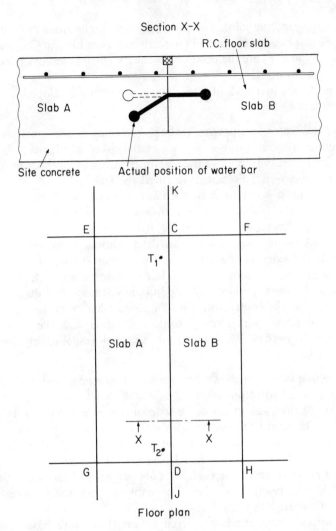

FIG. 9.8. Suggested method of repair of defective joint in ground slab.

a typical example will be given to illustrate what the author considers are important principles.

Figure 9.8 shows the situation as finally ascertained after lengthy investigations following the failure of a sewage tank to pass the prescribed water test. The displaced water bar was located by careful drilling alongside the joint. There were welded connections at C and D where the displaced water bar CD joined water bars EF and CK, and DJ and GH. T_1 and T_2 are the positions of the two drill holes which located the displaced water bar.

The problems involved, and the line of argument developed, may be stated as follows:

1. As the water was passing through the joint, it was obvious that neither the sealant nor the water bar were satisfactory. It was relatively easy to remove the sealant (which was a rubber bitumen) and replace it. This may have cured the leak.
2. However, it was known that sealants are not long lived materials, particularly rubber bitumens and polysulphides. Although neoprene has a much longer life, it is also likely to deteriorate in the course of time. As soon as deterioration starts and/or bond with the sides of the sealing groove begins to fail, leakage will recommence because the water bar is in some way defective. The water bar was specified as an additional safeguard, but due to some failure by the contractor, it no longer fulfilled its purpose.
3. The contract was already behind schedule and the tank was urgently needed. The contractor accepted responsibility for the remedial work.

The real question then was, what was a practical and reasonably satisfactory method of repair?

One suggestion was to cut out a strip of concrete about 1·00 m wide parallel to the joint CD. Another idea was to completely remove bays A and B, but this immediately raised the question of how to deal with the water bars in the joints EG, FH, EF and GH. A further suggestion was to try to pressure grout the defective concrete around the water bar in bay A. This was rejected as it would probably result in the grouting-in of part of the underfloor drainage.

The author's suggestions are given below; these were based on the principle that as little concrete as possible should be cut out.

(a) The existing sealant should be cut out and the sealing groove deepened by about 15 mm.
(b) Both sides of the sealing groove should be reformed by first cutting the concrete with a saw to widen the joint, and then trueing-up the sides with an epoxide mortar.
(c) After the reformed groove had been carefully cleaned out, a cold-

cured epoxide resin should be inserted to a depth of about 15 mm. This resin should possess low viscosity, flexibility and should bond to the damp concrete sides of the groove. The reason for providing this under layer of resin was that the final (top) sealant described in (e) below, would not bond to the existing sealants in the adjacent joints.

(d) After completion of (c), a cellular neoprene jointing strip should be inserted in the groove above the epoxide resin under layer. The strip would be bonded to the concrete sides of the groove by a suitable primer. It would not be bonded to the epoxy under layer.

(e) For a distance of 600 mm each side of the joint and for its full length, the surface of the concrete should be prepared by mechanical scabbler, grit blasting or high velocity water jets, so as to lightly expose the coarse aggregate to a depth of about 5 mm.

(f) An epoxide resin mortar should be laid on the prepared surface of the concrete to a minimum thickness of 5 mm, the resin to be specially formulated to bond to damp concrete.

The work was carried out generally as described, and no further leakage was observed. It was completed in five working days, compared with an estimated minimum of three weeks for any of the alternative schemes. The formation of additional joints, with corresponding hazards of leakage, was thus avoided.

9.4.3 General Repairs and Lining of Water-retaining Structures

While the principles of chemical attack on concrete are dealt with in Chapter 2, there are many borderline cases of mild chemical attack on concrete in service reservoirs due to certain aggressive characteristics of the raw or partially treated water. This occurs particularly where the water is derived from an upland gathering ground.

Such waters are characterised by:

Low total dissolved solids (TDS)
Low total hardness
Carbon dioxide in solution (aggressive, or free CO_2)
Low pH
Low alkalinity
Organic and other acids in solution
Sometimes a negative Langelier Index

Occasionally, the acids include sulphuric derived from the breakdown of sulphur compounds in peat and marshy ground by bacteria. The presence of sulphuric or other mineral acids will greatly increase the severity of attack. It is often found that the pH varies over a range from about 4·0 to 6·5. In some parts of the country there is a significant drop in pH after heavy rainfall.

While it is customary in some cases to treat the raw water before it enters the reservoir by filtration through limestone chippings and/or addition of lime, with sudden variations in the acid content of the water the pH correction may not always be as successful as one would desire.

These facts are mentioned because they have an obvious bearing on the assessment of any damage done and on the method of repair.

The pH value itself has no direct relationship to the amount of acid present; its potentional for attack can only be determined when the type of acid causing the low pH has been established by chemical analysis.

Water with a pH in the range of 7·0–8·0 is alkaline but this does not necessarily mean that it will not attack Portland cement concrete. Under certain conditions such water can attack concrete due to its potential in dissolving out the calcium hydroxide (lime). The lime dissolving power of such waters depends on the amount of dissolved carbon dioxide, the hardness expressed as calcium carbonate, the alkalinity measured by methyl orange and expressed as calcium carbonate, the temperature, and the original pH of the water, and total dissolved solids.

The Langelier Index, or calcium carbonate saturation index, can be used to determine whether the water is lime dissolving (denoted by a negative value to the Index) or lime depositing (denoted by a positive value to the Index).

The saturation pH (pH_s) is calculated from the total dissolved solids, the temperature (°C), calcium hardness as $CaCO_3$ and the methyl orange alkalinity as $CaCO_3$. The saturation pH is then subtracted from the original pH and the result is the Langelier Index of the water. For further information on the Langelier Index reference should be made to the book, *Water Treatment for Industrial and Other Uses*, by E. Nordell, and a paper by W. F. Langelier in the *Journal of the American Water Works Association* in 1936; both are listed in the Bibliography at the end of this chapter. Unfortunately there are wide differences of opinion among chemists on the aggressive effects of waters with different negative values of the Langelier Index.

The only specific information the author has seen on this important matter is a draft prepared by the International Organization for

FIG. 9.9. Concrete deeply etched by soft moorland water.

Standardization, Technical Committee ISO/TC77, Reaction of Asbestos-cement Pipes to Aggressive Waters. This draft recommends that a water should be considered as highly aggressive to asbestos–cement pipes if the Langelier Index is more negative than −2. It would be moderately aggressive with a Langelier Index in the range of 0·0 to −2·0.

The author considers it is prudent to have water checked for the Langelier Index in those cases where the normal chemical analysis suggests that the Index may be negative.

The Repair of Etched Surfaces
Attack on the concrete by soft acidic waters, and waters with a negative Langelier Index, normally show as etching of the concrete surface (see Figs. 9.9 and 9.10). The etching will vary in depth depending on the degree of attack which in turn will depend on the aggressiveness of the water and the chemical resistance of the concrete.

Provided the deterioration is confined to the surface of the concrete and there has been no significant corrosion of the reinforcement, the following measures can be adopted; the one actually selected will depend on the circumstances of each case:

Fɪɢ. 9.10. Deteriorated concrete channels carrying soft moorland water.

1. The surface of the concrete must be prepared by wire brushing or water jetting so as to provide a clean strong base on which the new coating will be applied. It is essential to do this preparatory work carefully so that the best possible bond is obtained.

2. On the prepared surface of the concrete, apply 2 thick brush coats of grout made with 2 parts ordinary Portland cement and 1 part styrene–butadiene latex by weight. The first coat must be well brushed in and the second coat should be at right angles to the first.

OR

3. Apply to the prepared surface of the concrete two coats of epoxide or polyurethane resin, to a minimum thickness of 0·3 mm. For drinking water reservoirs it is likely that polyurethane would not be acceptable as the coating must be non-toxic, non-tainting and should not support bacterial growth.

OR

4. Where the attack has been severe and there has been some penetration of water to the reinforcement resulting in limited corrosion, the application of a 50 mm thick layer of gunite,

reinforced with a light mesh, would be justified. Detailed information on gunite is given in a later section.

Bituminous compounds have in the past been used for this type of protection, but generally have not proved particularly durable. They do not bond well to damp concrete, and some may impart a phenol taste to the water.

The Complete Lining of Deteriorated Structures

Cases sometimes arise, particularly with water towers and similar elevated structures, where the provision of a complete lining is required in order to ensure a satisfactory standard of watertightness. For a complete lining, there are two basic alternatives: insitu material or preformed sheeting.

Insitu materials. These include cement grout composed of OPC and SBR latex, epoxide and polyurethane resins, and gunite. As previously mentioned, for structures holding drinking water, the coating must be non-toxic and non-tainting.

For the coatings, the correct preparation of the base concrete is essential. The author considers that it is most advisable for the supplier of the material to be responsible for all aspects of the work, i.e. preparation of the concrete, and supply and application of the coating, as this avoids divided responsibility.

Prior to the commercial use of organic polymers, such as epoxides and polyurethanes, mastic asphalt and bituminous emulsions were used for lining water-retaining structures. These latter two materials are now very seldom used; they have been replaced by Portland cement and styrene–butadiene slurry coats for mildly aggressive conditions and by polymer resins (certified as being suitable for use with drinking water when used in potable water reservoirs).

Latex grout: polymer emulsions and Portland cement. The treatment of the floor and walls of a water-retaining structure with a grout composed of two parts Portland cement and one part styrene–butadiene resin latex emulsion (SBR) by weight can be effective in sealing the surface of slightly porous concrete subjected to a low head of water. There are no figures available for maximum safe water pressure, but the author would not feel confident in the use of this material under a pressure exceeding 1·50 m head.

It is simple to apply and relatively cheap compared with more sophisticated coatings. Provided the limitations mentioned above are kept in mind, it can prove a useful solution to minor loss of water from small tanks and channels. There are a number of proprietary latexes on the market, but taking as an example the SBR latex 29 Y 40 manufactured by Revertex Ltd, and sold under proprietary names by various firms, the following is a guide to its use:

(i)   Mix: 2 kg ordinary Portland cement to 1 kg latex.
(ii) Coverage in two coat work: 5 m²/4 litres of grout.
(iii) The surface of the concrete must be clean and free of weak laitance.
(iv) Each coat must be well brushed in, and allowed to dry before the next one is applied. The second coat must be applied at right angles to the first.

Acrylic resin emulsions can also be used, but they tend to be more expensive.

In addition to the basic emulsions mentioned above there are a large number of proprietary materials on the market; the author feels that the most satisfactory results are likely to be obtained by employing firms who supply and apply the coatings.

Epoxide and polyurethane resin coatings. These materials are basically organic polymers and can be formulated to give the coatings a wide range of characteristics such as chemical resistance, bond strength, modulus of elasticity, viscosity, setting and curing time, ability to cure in very wet conditions including complete submergence, abrasion resistance and colour.

The client (or his professional adviser) should specify as clearly as possible the conditions under which the coating has to operate. If the coating has to possess considerable elasticity, this should be defined as accurately as possible. Such expressions as 'the coating must be capable of spanning hair cracks' should not be used as there is no general acceptance of the width of a hair crack.

This type of coating will bond very strongly to clean, good quality concrete, and the bond is an essential factor in the long term serviceability of the coating. When failure occurs it is normally bond failure at the interface with the base concrete. It is essential that all weak laitance and any surface contamination such as oil, etc., be completely removed. It is better if the preparation of the concrete and the formulation, supply and application of the coating are all carried out by one firm as this avoids divided responsibility.

For the complete lining of a liquid retaining structure, a minimum thickness of 0·5 mm is recommended and this would normally require three coats. The resin can be applied by brush or airless spray. When properly executed it provides a completely jointless lining which is very durable.

As with all applied coatings which require to be bonded to the base concrete, the limiting factor may be the actual quality of the concrete. With a weak porous concrete the effective life of the lining may be disappointing, simply because the concrete is not strong enough to hold the coating in position. In such a case, reinforced gunite may be the best solution.

Reinforced gunite. Gunite is a pneumatically applied material, consisting of cement, aggregate and water. Depending on the maximum size of aggregate used and the grading, it can be considered either as a mortar or a concrete.

The advantage of this material for watertight linings compared with concrete, is that it does not require formwork, is self-compacting, and that except for large structures, or in those cases where the gunite is bonded to the existing structure, all the joints are monolithic. In addition to providing a watertight lining, reinforced gunite can be designed to strengthen an existing structure. This use of gunite has been described in Chapter 5.

Gunite is used more extensively in the USA, South Africa and Australia than in this country. However, in recent years its use in the UK for new swimming pool shells and the lining of deteriorated pools and other liquid-retaining structures, has increased.

Gunite work is usually offered as a 'Design and Construct' package deal type of contract by a few specialist firms. With certain aggregate gradings and the use of specialist 'gunning' equipment, equivalent 28-day cube strengths in excess of 40 N/mm^2 can be obtained compared with about 35 N/mm^2 for normal quality water-retaining concrete. The higher compressive strength is accompanied by higher tensile and bond strength as well as impact resistance.

When reinforced gunite is used to line a structure, the reinforcement can be pinned to the old walls and floor and the gunite forms a monolithic lining bonded to the existing structure. An alternative, which is usually adopted when the existing structure is found to be in poor structural condition (low strength concrete, numerous cracks, etc.), is to separate completely the new gunite lining from the old structure. The separation is generally formed by a membrane of 1200 gauge poly-ethylene sheeting (0·3 mm thick). The reinforced gunite lining then forms an independent structure, and would be designed as such.

If a bonded lining is used, then all movement joints in the existing structure should be carried through the gunite. When a liquid-retaining structure has cracked at locations other than along the line of joints, it means that these cracks may act as movement joints. This is the principal reason why an independent unbonded lining is adopted.

It was mentioned earlier in this section that unless the structure is very large, the gunite-independent lining would have no movement joints but only monolithic day-work type joints. A natural question, therefore, is why should provision have to be made in a concrete lining for thermal contraction (stress relief or contraction joints) while gunite appears to be satisfactory without any movement joints at all. It is not easy to give a completely clear, unambiguous answer to this, but the factors involved are:

(a) A reinforced gunite lining is usually appreciably thinner than a lining in reinforced concrete. The gunite lining has no formwork, resulting in far less build-up of heat from the hydrating cement, and consequently lower thermal stress, than concrete.

(b) Gunite linings are usually reinforced with heavy steel fabric and the amount of horizontal (distribution) steel is likely to be greater per unit thickness of wall or floor than would be provided for concrete.

(c) Good quality, high strength gunite, as placed, has a very low w/c ratio, about 0·33–0·35. Concrete used for lining a liquid-retaining structure would normally have a w/c ratio of about 0·48–0·50.

The above factors add up to the fact that there would be appreciably less thermal contraction in the gunite lining than in an insitu concrete lining, followed by less drying shrinkage, coupled with a heavier weight of reinforcement to control cracking.

However, cracking is not unknown in gunite linings, but is generally caused by inadequate precautions being taken to prevent rapid drying-out of the surface under adverse weather conditions (hot sun, strong winds). This rapid drying-out shows in the form of fine surface cracks. It is advisable for provision to be made for proper covering up of the gunite as the work proceeds.

The bulk density of high strength gunite is about 0·9 that of normal dense concrete, i.e. $0·9 \times 2380 = 2100 \ kg/m^3$.

The surface finish of gunite is generally rougher than that of good quality concrete. This may be considered a serious disadvantage when gunite is used to line potable water tanks as some water engineers consider the rough surface would encourage the growth of algal and fungoid growths. This can be partly overcome by finishing the gunite with a wood or steel float.

Preformed Sheeting

There are three basic methods of lining water-retaining structures with sheet material: fully bonded lining; partially bonded lining (spot bonding); and unbonded lining.

In the UK for new structures such as reservoirs and water towers a fully bonded lining is normally adopted. In recent years, the technique of using a loose 'bag' of PVC sheeting has been introduced for small swimming pools and it could be used for other structures in certain restricted circumstances.

Fully bonded and partially bonded linings. The basic principles for successful application of these two types of lining are very similar and therefore are considered together. A partially bonded lining is used where appreciable further movement in the structure is anticipated. Care is taken to ensure that the lining is left unbonded across all planes of movement and in this way, excessive build-up of stress in the lining material is avoided. Generally about 75% of the area of lining is bonded to the substrate.

The great advantage of a bonded lining compared with an unbonded one is that should the lining be damaged in a limited number of places, it is most unlikely that leakage would occur. The reason for this is that the liquid cannot travel behind the lining until the adhesive has become very deteriorated. The adhesives used are water-resistant, but it must be admitted that if water does penetrate through the membrane, in the course of time the adhesive may suffer deterioration. The degradation of the adhesive and consequent loss of bond will show itself by the bulging of the lining and this should be detected in the course of normal maintenance inspections before the water is able to penetrate through the structure.

The disadvantage is the careful preparation which is required of the base concrete; this preparatory work is detailed below. However, on balance, the author is of the opinion that for structures such as water towers, etc., the advantages of a fully or partially bonded lining far outweigh the disadvantages.

As with any material requiring good bond, careful preparation of the base concrete is essential. In many cases, linings of this type are applied to old structures which over the years have been coated with other materials (such as bitumen) to help improve watertightness. All such coatings which are unsound or blistered must be completely removed. It is important that the surface to which the sheeting will be bonded is strong and relatively smooth. A scabbled or grit blasted surface would not be suitable as the irregularities on the surface created by the pieces of coarse aggregate would cause local stress in the sheeting. Therefore if scabbling or similar is required to remove contamination a thin fully bonded coat of rendering must be applied to the base concrete. If the concrete is reasonably clean, but contains rough areas, these must be ground down.

The sheeting should be well bonded to the concrete with a special adhesive which is not water-sensitive and all seams must be lapped and solvent (cold) welded. An alternative is to use a special tape which is inserted between the sheets and then the joint is fused.

Figure 9.11 shows a previously leaking reservoir after lining with fully bonded sheeting.

It is important that the sheeting be continuous over the floor, and carried up columns and walls to at least 300 mm above top water level. If the sheeting is finished below water level, then it is likely that continuous immersion of the unprotected edges over a long period of time will result in deterioration of the adhesive, causing loss of bond and allowing water to penetrate behind the lining. Even so, the top exposed edge of the sheeting needs careful finishing to avoid ingress of condensation as this is always present in tanks holding liquid at ambient temperature. Reservoirs and water towers contain a number of pipes, both inlet and outlet, and the sheeting must be carefully cut and fitted around these and all joints lapped and solvent welded. Any joints in the structure need special attention and detailing so as to prevent build-up of stress in the lining.

The lining of tanks with flexible sheeting is highly specialised and should only be entrusted to an experienced firm who should be required to give a guarantee of satisfactory performance for not less than 10 years.

FIG. 9.11. Inside of water reservoir lined with fully bonded sheeting (courtesy: Gunac Ltd).

Preformed sheeting is fairly easily damaged by small metal tools, hobnail boots, etc., and therefore special precautions have to be taken when carrying out inspections, maintenance and cleaning work.

Unbonded lining. The use of an unbonded lining of flexible sheet material for the waterproofing of water-retaining structures appears to have started with small private swimming pools. The material generally used is polyvinylchloride (PVC) (also known as vinyl), and the sheets vary in thickness from 0·80 to 1·5 mm. It is unlikely that PVC would be accepted for lining potable water tanks due to possible leaching of chemicals from the PVC. Chlorinated polyethylene is also used.

The advantages claimed for the use of unbonded lining are that movement of the structure does not induce stress in the lining, and that provided the surface on which it is laid is reasonably smooth, no special preparation is required. Also, with open structures (not roofed), the

Fɪɢ. 9.12. Deteriorated concrete tank holding industrial effluent (courtesy: Cement Gun Co. Ltd).

Fɪɢ. 9.13. View of tank in Fig. 9.12 during repair with reinforced gunite (courtesy: Cement Gun Co. Ltd).

FIG. 9.14. View of tank in Figs. 9.12 and 9.13 after completion of repair (courtesy: Cement Gun Co. Ltd).

sheeting can be laid in almost any weather. The author has seen this method used on roofs and for lining small swimming pools.

While the horizontal area which can be covered by unbonded sheeting is virtually unlimited, the vertical height must clearly be limited, probably to a maximum of about 1·5 m, although the author has not seen any figure quoted. With any form of complete lining, it is important that it should not be subjected to back-pressure from ground water when the structure is empty as there would be a danger that the sheeting would be forced away from the substrate. In the case of unbonded linings it is generally considered essential that all necessary precautions be taken to prevent any back-pressure developing, however slight. The weight (pressure) of the contained liquid moulds the sheeting to the exact shape of the structure, and if at a later date the lining is forced out of position by pressure from behind, it is unlikely ever to return to its previous position and shape.

The External Use of Reinforced Gunite to Strengthen Liquid-retaining Structures

A previous section has dealt with lining liquid-retaining structures with reinforced gunite. This type of lining is usually applied to below-ground structures. For structures which are above ground, such as many of the tanks at sewage treatment works, sludge and settling tanks for industrial effluent, and water towers, it is sometimes necessary to strengthen the tanks from outside. Reinforced gunite is very suitable for this work.

The gunite coating, reinforced as required to provide the necessary additional strength, is fully bonded to the base concrete. The existing concrete must be carefully prepared to receive the gunite. It is good practice to cut away all spalled and defective concrete, remove rust from reinforcement, cut out and repair all major cracks, and thoroughly clean the surface of the concrete by grit blasting or high velocity water jetting. The new reinforcement is then securely pinned to the old concrete and gunning is commenced.

Figure 9.12 shows a badly deteriorated concrete settling tank used for industrial effluent. Figure 9.13 shows the same tank being prepared for guniting. Figure 9.14 shows the same tank after completion of external structural gunite.

9.5 REPAIRS WHERE LEAKAGE IS INWARDS INTO THE STRUCTURE

So far in this chapter consideration has been given to the repair of liquid-retaining structures where the leakage is outwards from the structure. Leakage or seepage into the structure can take place with both liquid-retaining and liquid-excluding structures. In the former type the infiltration would only occur when the pressure outside exceeded the pressure inside, in other words when the structure is partially or completely empty. When infiltration does occur, then this can be very difficult to cure completely, because it is usually impossible to carry out the repair from the water (outside) face.

9.5.1 Repairing Leaks in Basements

Consideration of this problem includes basement structures in which special electronic equipment is fixed and no moisture penetration can be tolerated. In cases of this category, it is usual to tank the structure, i.e. provide a complete waterproof membrane under the whole of the floor and carry it up the external walls to above ground level. Even so, the problem of ensuring that such a large area of membrane is completely watertight and remains so during the construction of the structural floor and walls and for the lifetime of the building is a formidable one, and is not always successful.

It is obvious that at the design stage, detailed consideration should be given to the standard of watertightness required, bearing in mind that good quality, well compacted concrete is watertight but not vapourtight. The standard required for the sump of a sewage pumping station would be different to that of an underground car park, which in turn would differ from a basement used for storage of materials liable to be damaged by high relative humidity.

When the specified standard of watertightness has not been achieved and repairs are required, the problem is to decide how these repairs should be carried out, taking into account the circumstances of each case as far as they are known. The words 'as far as they are known' are important because in most cases nothing is known for certain about the actual condition (impermeability) of the concrete which is behind the inner face of the wall. All that is known is that water penetration is occurring; this may be along joint lines or it may be in random areas of the floor and walls.

In the case of penetration on joint lines, this may be due to a fractional opening of a 'monolithic' joint which does not have a water bar and/or sealing groove on the outside. If there is a water bar then the seepage indicates either some displacement of the water bar or honeycombed (undercompacted) concrete around the bar. In the random areas, this is due to undercompacted concrete.

What is emphasised here is that the extent of the undercompaction (or degree of porosity) is not known, unless cores are taken or an ultrasonic pulse velocity survey is carried out. It is the author's experience that it is very unusual for either of these two methods of investigation to be adopted. It is general practice to stop the leaks by sealing the inside face of the floor or wall. There are a considerable number of proprietary materials on the market in this country, the Continent, USA, etc., which, when properly used, will seal off the inflow of water. Most of these materials are fairly new inasmuch as they have been developed over the last 15–20 years. Their durability in terms of the normal lifetime of a reinforced concrete structure is therefore not known for certain. This statement is not intended to suggest in any way that these or other new materials should not be used. If this attitude were adopted, there would be no progress at all in the development of new materials and techniques.

Questions are sometimes asked whether the water penetration and consequent repairs will result in increased maintenance costs. It is the author's opinion that it is reasonable to assume that some additional maintenance expenditure will be required compared with the same structure if no water penetration had occurred at all. However, it would be quite unrealistic to assume that a large reinforced concrete basement can be constructed in a subsoil which contains a water table, without any water penetration at all. It can be done in theory, but not in practice unless the basement is tanked. From this a fundamental question arises, namely, what effect will water penetration have on the long term durability of the structure from the point of view of structural stability and maintenance. The author has already given his opinion on maintenance costs, but he feels that detailed consideration of the effect of water penetration would be useful.

The reinforced concrete walls of basements are either (a) load-bearing or (b) panels spanning between reinforced concrete columns. The floor slab is usually uniformly supported on the ground with additional reinforcement to take any uplift due to water pressure, or is a suspended slab, also with additional reinforcement. The first matter to be considered is whether the ground water will attack the concrete itself. Some ground waters are mildly aggressive to Portland cement concrete, but not sufficiently so as to cause any significant attack on an adequate thickness of high quality, dense, impermeable concrete. When sulphate-resisting Portland cement is used, it is always emphasised that a good quality well-compacted concrete is also required, so that the sulphate-resisting properties of the cement will operate to the best advantage.

It has been stressed several times before in this book that steel embedded in Portland cement concrete is protected from corrosion by the intense alkalinity of the cement paste. In other words the steel is passivated. Unless this passivation is broken down either by a reduction in the alkaline environment provided by the cement paste or by other factors such as the presence of chloride ions, corrosion of the steel will not occur.

If the leak is sealed off on the inside face of the wall or floor, this does not prevent water entering the concrete from the other side, but it does stop any flow-through. This prevention of flow and the establishment of static conditions is more important than may appear at first sight. Concrete subjected to continuous water pressure will in the course of time become saturated. The rate at which the water will penetrate the concrete will depend on the permeability of the concrete. With high quality, dense, well compacted concrete the permeability rate will be very low. This very slow passage of water into the concrete will not, as far as is known at present, result in deterioration of either the concrete or the steel reinforcement, unless the water contains aggressive chemicals in solution. Also, for corrosion to occur, a supply of oxygen is required at the surface of the steel. The amount of oxygen present below ground level is much reduced compared with the open air.

Clearly, the actual amount of water which penetrates into the concrete is an important factor. Unfortunately in the type of structure considered here, this factor is not known. If for example, the concrete in the outer part of the wall or lower part of the floor slab were very porous, there may be so much water penetration as to reduce the alkalinity of the cement paste below the level required for effective passivity and then corrosion of the outer (or lower) bars may occur. However, unless the areas of seepage were considerable, it is unlikely that the strength of the wall or floor as a whole would be affected to any significant degree.

From this brief discussion it is obvious that each case of water penetration must be considered on its merits. Surface sealing of walls and floors on the inside face, remote from the point of entry of the water, has been successful in providing a reasonably dry basement. It can therefore be considered as an established and accepted method of repair. An alternative or additional method is pressure grouting, which has been briefly described earlier in this chapter. It cannot be relied upon to completely seal off areas of infiltration, but if carried out by experienced contractors, it will greatly reduce the penetration of water. It has the advantage that the grout will penetrate into the sections of concrete which are honeycombed, thus providing direct protection to the reinforcement in that part of the wall or floor.

It is the author's opinion that a specification for a new structure of the type considered here should contain clear directions on how any necessary repairs to prevent ingress of water should be carried out. Such a specification may require both pressure grouting and surface sealing when the wet area or amount of moisture penetration exceeds stated figures as this would help to ensure long term durability.

9.5.2 Repairing Leaks in Roof Slabs of Water-retaining Structures
In some structures, minor leaks in the roof may not be viewed with much concern apart from the danger of corrosion of the reinforcement. However, in the case of drinking water reservoirs any leak is a potential source of contamination. Joints and cracks in the concrete slab are likely to be the principal cause of leakage. Porous concrete may contribute, but is seldom the main cause of the trouble.

It is now normal good practice to provide a waterproof membrane over the whole of the roof slab, but this was not the case some 30 years or so ago. The absence of the membrane is often accompanied with inadequate falls to the slab. This results in 'ponding' and may lead to the gradual saturation of the concrete. In the course of time the alkaline environment around the steel may be reduced to such a level that corrosion occurs.

Where there are no definite leaks, but rust stains on the soffit of the slab, it is very difficult to decide whether the corrosion is due to the porosity of the concrete cover, i.e. the soffit concrete, or to penetration of water from above.

The earth cover, if any, must be removed, and the surface of the concrete thoroughly cleaned, with special reference to joints and cracks. In repairing joints and cracks, the major decision has to be taken on the method of repair, whether to use a rigid or flexible material. With uncovered, exposed slabs, the temperature range may be as much as 70°C (see BRE Digest 228, August 1979), while with a cover of 300–400 mm of earth, the range may not exceed 15°C. With a large temperature range, the joints and/or cracks are likely to open and close seasonally and therefore flexible sealants should be used. With a much smaller temperature range, many of the joints and cracks can be safely repaired with a rigid material, so as to 'lock' the joint or crack.

Many older structures have no purpose-made movement or partial movement joints in the roof slab. This often results in some of the construction joints opening and forming what in reality is a number of stress relief joints. Cracks are sometimes formed by the same cause. Where there is a definite leak through a joint, a practical way to effect the repair is to remove the whole of the existing sealant and replace it with new. If there is any inert filler in the joint this should also be renewed. The new sealant can be any of the materials described in Chapter 1, bearing in mind the characteristics of the various types. If a preformed gasket is selected, the width of the gasket must be wider than the groove into which it will be fixed, and the use of the correct 'oversize' is essential if a watertight joint is to be achieved. Figure 9.15 shows the use of Hypalon sheet and an insitu sealant. If a crack is sufficiently straight it should be possible to seal it with the method shown. Unfortunately cracks are seldom straight, and then the only practical method is to cut it out with the special tool (Fig. 6.2, p. 154).

In structures built in the 1920s and 1930s, it is sometimes found that the roof slabs are reinforced with XPM instead of round bar reinforce-

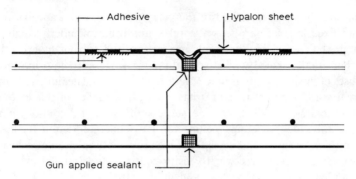

FIG. 9.15. Method of sealing contraction joint in roof slab.

ment as is the practice today. The author has found that quite severe corrosion of XPM can occur without spalling or cracking of the concrete. Rust stains are visible and sometimes the outline of the XPM can be clearly seen. When this happens the concrete should be removed to allow the XPM to be examined and an assessment made of its value as tensile reinforcement. If the XPM is seriously corroded, it is advisable to provide new reinforcement on the soffit of the slab, anchored into the beams and properly gunited in. At the same time a decision should be taken as to the cause of the corrosion, i.e. penetration of water from above or below. In practice it may not be possible to arrive at a clear cut answer to this, and in this event, it would be prudent to provide protection on the top surface and on the soffit of the slab.

If the suspected seepage is widespread, the provision of a new fully bonded waterproof insitu membrane of polyurethane or preformed sheeting would be justified. An alternative is the use of unbonded sheeting. The advantages of this latter method are that the sheets, being unbonded, are not subjected to strain due to movement of the roof slab; also, they can be laid in almost any weather.

However, if the unbonded sheets become damaged and water can pass through the holes or tears, then it will gradually flow over the roof slab until it finds a weak spot and then will penetrate the slab. With fully bonded sheets this will not occur as long as the adhesive remains intact.

Roofs of reservoirs are sometimes very exposed and it is essential that the unbonded membrane be held down against the suction which develops during periods of storm and strong wind. This is usually done by carefully spreading rounded (not angular) shingle on the surface to a depth of about 50 mm.

The sheeting is in fact fixed to the perimeter of the roof and is carried up and fixed against parapet walls, pipes and other members which pass through the roof slab. The sheets are lapped and solvent (cold) welded and then finally sealed with a special material. Detailed information on the use of insitu coatings and preformed bonded sheeting for lining reservoirs and similar structures has been given earlier in this chapter.

Other materials which can be successfully used to hold-down and protect unbonded sheeting, include no-fines concrete 75 mm thick, and 50 mm thick precast concrete slabs. These two materials can also be used to provide a protective cover to insitu coatings and bonded preformed sheeting, capable of taking foot and light traffic.

There will be cases where a membrane has been provided, but in spite of this, the roof leaks. The location of the defect(s) in the membrane can be very difficult if not impossible because it is extremely unlikely that points of visible leakage on the soffit of the slab will coincide with the defects in the membrane. The decision on the best method of repair will depend largely on the extent of the leakage. If the leakage is extensive probably the most satisfactory repair will include the removal of the existing membrane and its replacement by new material. At the same time the opportunity should be taken to seal the defects in the concrete slab. If the leakage is relatively small it can be repaired on the underside of the slab by one of the methods described in the previous section.

9.6 REPAIRS AND WATERPROOFING OF NEW TUNNELS AND PIPELINES

9.6.1 General

This Section deals in a general way with repairs and waterproofing which are sometimes needed to new tunnels and pipelines and arising from incidents during construction. It does not cover the much larger subject of the renovation of old sewers which may be brick lined tunnels and pipes of concrete or clayware. A few references on this latter subject are included in the Bibliography at the end of this chapter; the total amount of published material on sewer renovation which has appeared during the past ten years is quite staggering.

The materials and techniques which have to be used in the waterproofing of tunnels and pipelines are in many ways different to those adopted for waterproofing structures such as reservoirs and sewage tanks. Tunnels and pipelines must inevitably be waterproofed on the inside and against the inflow of water, while, as discussed earlier in this chapter, repairs, etc., to water-retaining structures can often be carried out on the water face.

Tunnels and large diameter pipelines are usually driven or laid through water-bearing ground and the inflow of water, particularly in the case of tunnels, can be very considerable. Most consulting engineers realise this and accept a certain amount of infiltration. Many papers have been written about tunnels and tunnelling, and also about main intercepting sewers, but it is rare to find this basic problem of watertightness discussed, and details given of the amount of infiltration accepted by the designers. An exception to this is the paper by Haswell on the Thames Cable Tunnel, read at a meeting of the Institution of Civil Engineers in London on 17 February, 1970.

In the case of tunnels, the amount of infiltration will depend on many factors including the pressure of the ground water, the type of lining, whether single or double skin, the materials and techniques used for caulking the joints between the segments and how leaks elsewhere are dealt with. Haswell in his paper quotes acceptance figures of 5·5 litres/m^2/day, and 32 litres/m^2/day in different sections of the tunnel.

In addition to the tunnel lining it is usual to pressure grout behind the lining as work proceeds. The joints in cast iron segments are usually caulked in lead while with precast concrete segments, the caulking material is normally a cement based compound.

None of these are 100% effective and so additional waterproofing is required on a certain percentage of the joints. In addition, there is always some leakage at intermediate places. To seal these leaks against the inflow of water, ultrarapid setting compounds have to be used. There are many proprietary materials on the market. The older ones were usually based on Portland cement with an admixture of gauging liquid to promote an almost instantaneous set. Portland cement and HAC in about equal proportions will give a flash set. In recent years organic polymers, such as epoxide, polyurethane, polyester, and acrylic and styrene–butadiene resins, have been introduced for this waterproofing work. Sometimes they are used in addition to the older materials. The formulators of some of the polymers can produce compounds with special characteristics tailor-made for specific site conditions.

Underground pipelines for main sewers are almost entirely reinforced concrete; asbestos-cement was also used in the UK for diameters up to about 1·20 m. Asbestos-cement pipes up to 2·0 m diameter are made and used on the Continent and the USA. Both types of pipeline have flexible

FIG. 9.16. Infiltration through joint in concrete pipeline   (courtesy: Colebrand Ltd).

joints formed with rubber rings. When correctly installed these form a watertight joint, but if the rings are displaced during pipelaying or grit gets between the rings and the pipe, leakage will occur. It is usually at joints in the pipeline and the junction between the pipes and manholes that infiltration is most likely to occur.

With large diameter pipes, 900 mm and over, men can enter the line and inspect and repair the joints. When there is a considerable inflow of water, repair can be difficult and time consuming. It is impossible to remove the jointing rings and to correct their position. The joint has therefore to be made watertight by the insertion of a sealing compound into the narrow space between the end of the spigot of one pipe and the inside of the back of the socket of the next. Sometimes due to excessive ground movement, or incorrect laying, this space may be extremely narrow in one half of the joint and very wide in the other. Such conditions require the concrete to be cut back so that there is an adequate space for the sealant.

In one particular job where about one thousand such joints had to be made watertight, the prescribed dimensions of the sealing groove were 30 mm wide and 40 mm deep. The main items of work that had to be done were:

(a) The joints had to be cut out and prepared to the dimensions given above.
(b) The inflow of water then had to be completely sealed off by means of an ultra-rapid setting compound.
(c) On completion of (b), the joint surfaces had to be prepared to ensure maximum bond with the selected sealant.
(d) The sealant, a specially formulated flexible polyurethane compound, was inserted and trowelled off flush with the inside of the pipe.

Figures 9.16 and 9.17 show the infiltration through a joint, and work in progress to seal the joint.

With pipelines having diameters less than about 900 mm, the tracing of infiltration points and other defects can only be carried out by closed circuit television. However, it is not possible to assess by photographs just how much damage has been done to a pipeline by chemical attack, in terms of the thickness and soundness of the remaining concrete.

9.6.2 Pipelines Constructed by Jacking

During the past 20 years, the use of pipe jacking techniques to lay large diameter pipes without excavating trenches has increased considerably. In common with all construction methods, it has advantages and disadvantages. A disadvantage is that damage can be caused to the pipes during the jacking; this usually arises from the pipes diverting from the prescribed line and level due mainly to changes in ground conditions.

The damage thus caused is usually at the joints, and unless it occurs near the leading pipes at the end of a thrust, the only practical solution is to repair the pipes from the inside. In pipes manufactured for jacking, the joints are normally of the spigot and socket type and are within the thickness of the pipe wall. The damage is mainly spalling and cracking at or close to the joints.

Where the spalling is minor it can be repaired using an epoxide mortar, particularly at the arrises of joints. For larger areas of damage, the better solution is to use a cement/sand mortar or a cement-rich,

FIG. 9.17. View of joint in Fig. 9.16 after initial sealing (courtesy: Colebrand Ltd).

10 mm coarse aggregate concrete, both being modified with the addition of 10 litres of styrene–butadiene emulsion to 50 kg cement. The repair technique should be as previously described in this book. Cracks can be dealt with as previously described and as appropriate in the circumstances.

There is sometimes a tendency for resident engineers to order the concrete in the pipe wall to be cut back behind the reinforcement. This is not necessary unless the concrete is damaged or otherwise unsound; no useful purpose will be served by removing concrete which probably has a compressive strength of 50–60 N/mm², and a water/cement ratio of about 0·35.

During jacking, particularly on a curve, the gap at the joints which is there to allow for some degree of flexibility, closes up around part of the circumference of the pipe. It is important that a reasonable gap width be maintained for the complete circumference of the pipe, of say 10 mm.

Any significant reduction in this width should be rectified by careful cutting with a disc. The new exposed surface should be given a brush coat of epoxy or polyurethane resin.

9.7 REPAIRS TO CONCRETE DAMAGED BY CAVITATION

Some brief information on the causes of cavitation damage to concrete was given in Chapter 2, Section 2.3.4. Generally cavitation damage occurs on the surface of spillways, aprons, energy dissipating basins, penstocks, syphons and tunnels which carry high velocity water. There is no agreed figure for the critical velocity below which cavitation damage will not occur, but a figure of about 15 m/s has been quoted. Damage to concrete caused by cavitation can be fairly easily distinguished from normal erosion, as the damaged areas have a jagged appearance while erosion by grit-laden water results in a comparatively smooth and/or rounded surface.

From published information on the subject of repairs to cavitation damage, it appears that unless the basic cause of the trouble can be removed, and this is very seldom possible, no repair method yet devised will guarantee lasting success. It is hoped that the following suggestions will prove useful.

1. All defective and damaged concrete must be removed. It is essential that the new concrete or mortar should bond very strongly to a sound high strength base. This may involve cutting away a considerable quantity of concrete. The new concrete or mortar must be high strength; as stated previously in this book, attempts to bond a high strength topping to a weaker base will almost certainly result in failure.

2. The concrete used for repair should have a minimum cement content of 400 kg/m^3, unless the thickness of the new concrete exceeds about 750 mm when the cement content can be reduced to 360 kg/m^3, a maximum w/c ratio of 0·40 and the aggregates should be a high quality crushed rock or flint gravel with a well graded clean concreting sand. Full compaction of the concrete, followed by adequate curing, is essential. Special care must be taken to provide as smooth a surface as possible, and there must be no lipping with adjacent concrete. Any high spots which remain should be ground down, and depressions filled with a fine epoxide mortar.

3. The junction of the old and new work is particularly vulnerable as a fine shrinkage crack usually appears around the perimeter of the new concrete. It is advisable to wire brush a band of a total width of about 300 mm (150 mm on the new concrete and 150 mm on the old) to remove all weak laitance, and then to apply two coats of epoxide resin.

4. Usually the repairs have to be carried out as quickly as possible, but it is important that speed should not be allowed to adversely affect the quality of the work. Consideration can be given to the use of ultrarapid hardening Portland cement or HAC. In both cases the minimum cement content should be the same as for ordinary Portland (as recommended above), but with HAC, the maximum w/c ratio should be 0·40. Advice on the use of both these cements should be obtained from the manufacturers; the reader should also refer to Chapter 1.

5. If the damaged area is extensive, but of shallow depth, consideration can be given to the use of high strength gunite having a minimum thickness of 75 mm. Depending on the area, it is usual for the gunite to be reinforced with a light galvanised fabric. The gunite should be specified to have a minimum equivalent cube strength at 28 days of 50 N/mm². Gunite work is highly specialised and should only be entrusted to an experienced firm.

6. Experience in the USA on repairs of spillways and stilling basins has shown that the provision of a sprinkle metallic finish to high quality concrete gives improved resistance to abrasion and cavitation. This type of finish is referred to in more detail in the next section.

7. Recent work in the USA suggests that polymerised fibre reinforced concrete may be particularly resistant to both cavitation and abrasion. Severe damage was caused to the spillway of the Dworshak dam in Idaho and after extensive laboratory research, the Corps of Engineers decided to use fibre reinforced concrete which is polymerised insitu for the repair of the most badly damaged areas. Brief information on polymerised concrete is given in Chapter 1.

9.8 REPAIRS TO CONCRETE DAMAGED BY WATER CONTAINING GRIT

The factors which are important in determining whether fast flowing water will or will not abrade the surface of concrete were summarised in

Chapter 2. Of the four listed factors, probably the most important is the quality of the concrete. The appearance of concrete damaged by grit-laden water is quite different to that caused by cavitation. In channels and pipes the wear may be confined to bends, or it may extend along the invert for comparatively long lengths. Really good quality concrete is surprisingly resistant to abrasion. Surface water sewers laid at very steep gradients down the slopes of Mount Carmel in Haifa showed no sign of damage after more than 15 years. There was only flow in these pipelines during the winter (November to March) but during that period of five months the average annual rainfall was about 600 mm.

On the other hand, in the UK, large insitu reinforced concrete channels used to convey sugar beet mixed with earth, stone and grit, were worn away to a depth of 50–75 mm in about two years The abrasive conditions were very severe and, in addition, the water in the channels contained sugar in solution which is chemically aggressive to Portland cement concrete.

If it is established that the quality of the base concrete is satisfactory and erosion has only occurred in a number of isolated places, then these can be repaired by careful patching with high strength concrete. It is advisable for these patches to be cut out with a saw or high velocity water jet, so as to provide a thickness of new concrete not less than 75 mm. If this minimum thickness cannot be provided, then a high quality cement/sand mortar containing an SBR latex should be used. Even so the thickness should not be less than 30 mm. For thinner patches, an epoxide resin mortar should be used.

Regarding the execution of the work, the general recommendations for the repair of cavitation damage will apply, with the exception that the provision of a very smooth surface is not usually required. However, this will depend on the hydraulic requirements of the structure.

Mention was made under the repair of cavitation damage to the use of a metallic sprinkle finish. This type of finish to industrial floors where high abrasion resistance is required has been described in Chapter 6. It is best applied to an adequate thickness of high strength insitu concrete while the concrete is still in the plastic state. The sprinkle finish consists of Portland cement, a metallic aggregate and an admixture, and is spread over the base concrete a few hours after compaction and finishing. The time factor is very important and can only be determined by experience. This thin topping is finished by careful hand trowelling, and then the newly laid area is cured in the usual way. When ferrous metal is used, brown stains will appear; this is caused by the rusting of the fine particles of iron in the surface layer. The rusting has no adverse effect on the

FIG. 9.18. Equipment for laying abrasion resistant invert in concrete pipes
(courtesy: Colebrand Ltd).

durability of the material. The thickness of the metallic finish depends
largely on the weight of metal used. With about 5 kg/m² of metal, the
thickness would be about 4 mm. For very severe conditions on industrial
floors, 45 kg/m² of metal with a thickness of 12 mm is sometimes used.
These metallic sprinkle finishes are applied by specialist firms, and can be
laid on slopes up to about 45° to the horizontal.

The type of repair so far described can only be carried out in locations
where there is access for the workmen. This means, in the case of pipes, a
minimum diameter of 1000 mm. For pipes smaller than this, special
equipment and techniques have to be used. Equipment has been developed and used for providing a special abrasion resistant invert to
asbestos-cement and concrete pipes. The material used consists of calcined bauxite mixed with a specially formulated epoxide resin. The
bauxite is expensive, and it is likely that carefully selected, graded and
washed flint would be satisfactory in many cases. Figure 9.18 shows such
an invert being laid in concrete pipes.

BIBLIOGRAPHY

AMERICAN CONCRETE INSTITUTE, *Manual of Concrete Practice*, Part 1, Erosion resistance of concrete in hydraulic structures, ACI, Detroit, 1968, pp. 210–13.

AMERICAN CONCRETE INSTITUTE, *Guide to the use of waterproofing, damproofing, protective and decorative barrier systems for concrete*, ACI Committee 515; ref. 515.1R-79, 1979, p. 41.

AMERICAN CONCRETE INSTITUTE, *Guide to joint sealants for concrete structures*, ACI Committee 504; ref. 504R-77, 1977, p. 58.

AMERICAN CONCRETE INSTITUTE, *Joint sealing and bearing systems for concrete structures*, Proc. of World Congress; ref. SP-70, 1981, 2 vols.

ARCHITECTS JOURNAL INFORMATION LIBRARY, Information Sheet 8, *Basements and solid ground floors*, Sept. 1976, pp. 413–18.

BRITISH STANDARDS INSTITUTION, *British Standard Code of Practice CP 2005:1968, Sewerage* (now under major revision).

BRITISH STANDARDS INSTITUTION, BS 5337:1976, *Code of Practice for the structural use of concrete for retaining aqueous liquids.*

BRITISH STANDARDS INSTITUTION, *Structural use of concrete*, BS 8100, Parts 1 and 2:1985.

BUILDING RESEARCH ESTABLISHMENT, Estimation of thermal and moisture movements and stresses, Part 2, Digest 228, August 1979, p. 8.

CLARK, R. R., Bonneville dam stilling basin, *J. Am. Conc. Inst.*, **52** (April 1956), 821–37.

CONSTRUCTION INDUSTRY RESEARCH AND INFORMATION ASSOCIATION (CIRIA), *Guide to the design of waterproof basements*; Guide No. 5, Feb. 1978, p. 38.

CREASY, L. R. AND WHITE, L. S., Water excluding structures. *Proc. ICE*, Paper 6395/6, Vol, 14, Sept/Dec. 1969, pp. 31–43.

DEPARTMENT OF THE ENVIRONMENT, *Watertight basements*, Part 2, Advisory Leaflet 52, HMSO, 1971, p. 4.

DEPARTMENT OF THE ENVIRONMENT AND NATIONAL WATER COUNCIL, *Sewers and water mains—a national assessment*. National Water Council, London, June 1977, p. 34.

GIFFORD, E. W. H. AND BUTLER, A. A. W., Elephant and Castle shopping centre. *Proc. ICE*, Paper 6906, Vol. 33, Jan/April 1966, pp. 93–119.

GREEN, J. K. AND PERKINS, P. H., *Concrete Liquid Retaining Structures*, Applied Science Publishers, Barking, Essex, Dec. 1980, p. 355.

KENN, M. J., *Factors influencing the erosion of concrete by cavitation*. Construction Industry Research and Information Association (CIRIA), London, Technical Note, 1968, p. 15.

LANGELIER, W. F., Chemical equilibrium in water treatment, *Journal AWWA*, **38**(2), February 1946, p. 169.

NORDELL, E., *Water Treatment for Industrial and Other Uses*, 2nd ed., Van Nostrand Reinhold, New York and Chapman and Hall, London, 1961.

PERKINS, P. H., *Swimming Pools*, 2nd ed. Applied Science Publishers, London, 1978, p. 398.

PERKINS, P. H., Corrosion problems in sewerage structures, Paper no. 12, *International Conference on Restoration of Sewerage Systems*, London, June 1981, Institution of Civil Engineers.

PORTLAND CEMENT ASSOCIATION, Concrete basements for residential and light building construction, *Concrete Information*, IS 208 01B, 1980, p. 7.

SCHRADER, E. K., AND MUNCH, A. V., Fibrous concrete repair of cavitation damage *Proc. ASCE Journal*, Construction Div., 102(CO2), June 1976, 385–402.

STRICKLAND, L., *Sewer renovation*, Technical Report TR87, Water Research Centre, Swindon, Sept. 1978. p. 52.

VICKERS, A. J., FRANCIS, J. R. D. AND GRANT, A. W., *Erosion of sewers and drains*, CIRIA Research Report No. 14, London, 1968, p. 20.

Chapter 10

Repairs to Concrete Marine Structures

It is generally acknowledged that a marine structure is located in a hostile environment and the conditions of exposure are classified as very severe (BS 8110). Experience shows that such structures are more liable to damage and deterioration than the majority of land-based structures.

Deterioration and damage can occur to a concrete marine structure in many ways which depend largely on its geographical location and site conditions, as well as its design, method of construction and the quality of the concrete. In areas such as the Red Sea and Persian Gulf, the sea water is appreciably more saline than in the more temperate zones. In the far North and far South the sea freezes for part of the year and the sub-zero temperatures can cause physical disintegration of the concrete. When the moisture in the surface layers turns into ice, the consequent expansion disrupts the concrete. Generally, the most vulnerable part of a structure is the section within the splash zone, that is from low tide to a certain variable height above high tide. The overall height of the splash zone depends on the degree of exposure, the weather conditions, and the tide range.

Site conditions also have a profound effect on the durability of concrete in this class of structure. Some sites are comparatively sheltered while others are exposed to heavy seas and gale force winds. Structures such as sea defence walls and groynes are sometimes subjected to severe abrasion by sand and shingle being dashed against the exposed faces. On the other hand, those parts of a structure which are permanently below low water level are in relatively uniform conditions of temperature, are not subject to alternate wetting and drying and are not exposed to the full fury of waves and wind.

Research into the durability of concrete in sea water has been carried out in a number of countries over many years, and reports on some of this work are included in the Bibliography at the end of this chapter.

Professor O. E. Gjorv of Norway has reported on concrete specimens immersed for periods of up to 30 years (Journal of the American Concrete Institute, January, 1971). These tests showed that Portland cements with relatively low C_3A (tricalcium aluminate) content (less than about 8%) were more durable than similar cements with higher C_3A contents. The tests also proved that HAC was as durable as the best Norwegian and German Portland cements. Even so, a complete understanding is still lacking of the reactions between Portland cement and the salts in solution in sea water.

In practical terms, the evidence from actual structures is conflicting. Some structures deteriorate more severely than can be satisfactorily explained, while others show unexpected and remarkable durability. Some of the steel reinforcement in the concrete Mulberry Harbours, constructed in 1943 for the Normandy landings, has been found in excellent condition after 30 years, with a cover of only 25 mm. It is clear that the quality of the concrete, in terms of cement content and impermeability, is the dominant factor, and not the thickness of the cover. This fact is confirmed by the durability (freedom from corrosion of the reinforcement) of concrete boats and ferro-cement boats, where the cover is often appreciably less than 25 mm.

The chemical constituents of sea water have been discussed in the section in Chapter 2, Section 2.5, dealing with the use of sea water for mixing concrete and mortar.

As far as chemical attack on the concrete is concerned, it is the sulphates in solution which may cause trouble. In Atlantic water, the sulphate concentration is about 2000 ppm (mg/l) and is largely magnesium sulphate, which is considered more aggressive than the sulphates of calcium and sodium in equal concentrations. In ground water, a concentration of 2000 ppm would require the use of sulphate-resisting Portland cement. However, it is well known that Portland cement concrete does not suffer from sulphate attack in normal sea water around the coasts of the UK, provided the concrete is high quality. This apparent anomaly has not been satisfactorily explained, but it is thought that the other salts present have an inhibiting effect on the reaction. It should be noted that the above remarks apply to 'normal' sea water. In some cases the water in estuaries and harbours is contaminated by sewage and trade wastes; also, if tidal currents are obstructed, there may be a build-up of salts in solution as well as aggressive organic compounds. In such cases chemical attack on the concrete can occur.

Except in extreme cases it is the author's experience that it is the quality of the concrete in terms of high cement content, low water/

cement ratio, and thorough compaction which will ensure a high resistance to sulphate attack, and sulphate-resisting cement is not
necessary.

In the mid 1970s, due to the decision to exploit the off-shore oil
resources in the North Sea, a research programme, sponsored by the
Department of Energy, was put in hand. It was named 'Concrete in the
Oceans' and its aim was to provide additional knowledge on the likely
long term performance of reinforced concrete oil production platforms,
some of which would be anchored into the sea bed at depths down to
about 250 m. This book does not deal with the problems of repair and
maintenance of these deep sea structures and is confined to the repair of
what may be termed 'normal' marine constructions, such as jetties, sea
walls, etc.

The author has found that three of the Reports are of particular
relevance to the problems of deterioration of normal type marine
structures, these are:

Report No. 2/11, Cracking and Corrosion, Aug. 1976, by A. W. Beeby.

Report No. 5,   Marine durability survey of the Tongues Sands
                  Tower, Taylor Woodrow Research Laboratories,
                  1980.

Report No. 6,   Fundamental mechanisms of corrosion of steel rein-
                  forcements in concrete immersed in sea water, by N.
                  J. M. Wilkins and P. F. Lawrence, 1980.

Readers are referred to these Reports for detailed information.

The Reports confirmed that the most severe conditions from the point
of view of deterioration occur in the splash zone; also, that reinforcement
embedded in concrete below low water level, i.e. permanently submerged,
corrodes very slowly in spite of the diffusion into the concrete of
chlorides. The passivity provided by the cement matrix is only reduced
slowly, and thereafter, active corrosion proceeds very slowly compared
with concrete containing similar concentration of chlorides exposed to the
air. The main reason for this is the considerable reduction in the amount
of oxygen present on the surface of the steel under permanently submerged conditions.

10.1 GENERAL PRINCIPLES OF REPAIR

A fundamental principle of the repair of any concrete marine structure is
that the concrete and mortar used must be of the highest quality in terms

of cement content, w/c ratio, compaction and impermeability. There will be, in addition, special requirements which will depend on the type of repair, its location, and site conditions.

The basic concrete mix should comply with the following:

Cement. This is generally ordinary or rapid hardening Portland cement. Sulphate-resisting Portland cement need only be used in special circumstances where the sulphate concentration and/or sea temperature is appreciably higher than in normal Atlantic water. HAC can be used with advantage in cases where a very high rate of gain of strength is required, such as concreting or guniting between the tides. The setting time is similar to that of Portland cements. Further information on HAC concrete is given in Chapter 1. Although long term tests on HAC concrete immersed in sea water have shown it to be as durable as Portland cement concrete it is advisable for the manufacturers to be consulted before a final decision is taken. There must not be leaching of alkalis from the Portland cement concrete into the HAC concrete.

Cement content. With Portland and HAC the cement content should not be less than 400 kg/m^3 of compacted concrete. In the case of mortar, the mix should not be leaner than 1 part of cement to 3 parts of well graded sand by weight.

Water/cement ratio. With Portland cements, this should not exceed 0·40, and with HAC, it should also not exceed 0·40.

Aggregates (from natural sources, complying with BS 882). In some parts of the world, the aggregates are contaminated with salt, often chlorides, and there may be insufficient fresh water available for adequate washing to remove the salt. As previously stated in this book, if the percentage of chlorides expressed as anhydrous calcium chloride exceeds 0·40% by weight of the cement, corrosion of the reinforcement may occur. Precautions can be taken as described in Chapter 2, in the section dealing with the use of sea water for mixing concrete. However, if the concentration is at, near, or a little above the limit of 0·40%, the cement content can be increased and this will have the effect of reducing the percentage.

Workability. The workability must be adequate to secure full compaction under the conditions of placing which will exist on site. This may require the use of a plasticiser with Portland cement concrete. There is seldom any need to use plasticisers with HAC due to the fact that it is more coarsely ground than Portland cement.

Admixtures. Admixtures, apart from workability aids, should generally not be used. With HAC no admixture should be used without the agreement of the manufacturers. In very cold climates, the use of air-entrained Portland cement concrete and mortar can be an advantage in combating the disintegrating effect of freezing temperatures on saturated concrete. Admixtures containing chlorides must not be used with HAC.

The term 'workability aids' is intended to include water-reducing admixtures, which in fact enable the prescribed workability to be maintained but with a reduced water/cement ratio, which has obvious advantages. The use of a suitable superplasticiser will enable a concrete with a low w/c ratio to be used and compacted (see also Section 10.3).

Cover to reinforcement. The relevant Code of Practice, BS 6349–Maritime Structures, recommends a minimum cover of 60 mm. The author considers that this depth of cover can result in shrinkage and thermal contraction cracking as it is known that the closer crack control steel is to the surface, the better. However, this can be overcome by special care in curing and the application of a sealing coat, both of which are discussed in Section 10.3. BS 8110, Part 1:1985, Table 3.4 shows a minimum cover to all reinforcement in concrete in a marine environment as 50 mm.

Other matters for consideration. When using cement/sand mortar for repairs, the author recommends the inclusion of SBR emulsion in the mix, and a cement/SBR slurry coat on the rebars immediately prior to the application of the mortar. When concrete has to be cast in formwork it is not practical to use the slurry on the rebars because it will set before the concrete can be placed.

Essentially, the methods and techniques of repair are similar to those in Chapters 4 and 5. Investigations should include the use of the cover meter and half-cell, as described in Chapter 3. Testing for concentration of chlorides and sulphates is essential.

The conditions of exposure are very severe and any defects in workmanship and/or materials will soon become apparent. There are many special problems associated with the execution of repairs to marine structures, including accessibility, work below water level and between the tides, and the effect of sudden storms on newly completed repairs.

Sea water is very aggressive to steel reinforcement and so the real problem is to ensure that rebars are properly protected in high quality, low permeability concrete.

10.2 REPAIRS BELOW LOW WATER LEVEL

In most cases repairs below low water level require relatively small quantities of material as the damage usually consists of cracked and spalled concrete in piles, piers, etc. However, it sometimes happens that fairly large volumes of concrete have to be placed under water to protect foundations against scour. There are three basic methods of placing concrete underwater: by tremie pipe or bottom opening skip; by bagged concrete; by grout injection of preplaced aggregate.

The following are brief notes setting out the most important factors involved in this type of work. The first, which applies to all three methods, is that work below water level is completely different to work on dry land. It is not a job for amateurs, and should only be entrusted to contractors with the necessary experience and equipment.

10.2.1 Placing by Tremie Pipe

The slump must be very high, 150–200 mm, which is virtually a collapse slump; at the same time the mix should be sufficiently cohesive so that it will not segregate during transporting and placing. The cement content must be high, not less than $400 \, kg/m^3$, and the w/c ratio should not exceed about 0·5. The use of a workability aid (plasticiser) is often necessary. The tremie pipe is usually 150 mm in diameter for 20 mm aggregate and 200 mm in diameter for 40 mm aggregate.

It is essential to organise the work in great detail so that a continuous supply of correctly designed concrete is assured. Once the concrete has started to flow, the end of the tremie pipe must remain submerged to a depth of about 400 mm in the concrete. An adequate number of tremie pipes should be provided; usually one tremie will cope with up to $30 \, m^2$. When the concrete has to be placed in more than one lift, the surface of the previous lift must be properly prepared by divers to remove all laitance, etc. As far as possible, the top surface of each lift should be horizontal.

10.2.2 Placing by Bottom-opening Skip

It is important that the skp should be filled to capacity with concrete and then lowered slowly through the water onto the concrete previously placed. The weight of the skip and its contents is sufficient to ensure that it sinks below the concrete surface. The skip is then gently raised and the concrete should flow out and into the surrounding plastic concrete. The mix proportions and very high slump are similar to those recommended for tremie work. The placing of the concrete is slower with a skip than a tremie.

Some engineers experienced in the use of both tremie pipes and bottom opening skips consider that the latter are usually more suitable for placing small volumes of concrete.

10.2.3 Bagged Concrete

Bagged concrete, according to the Report of the Concrete Society, No. 52.018: Underwater Concreting, is now only used for very minor and temporary works. However, the author's experience is that it can be very useful for protecting structures from physical damage and from the effects of scour; it can be used in both shallow and comparatively deep water.

The concrete should have normal mix proportions for marine work, namely 360–400 kg cement/m^3 of concrete. It is mixed fairly dry, with a w/c ratio of 0·35 or less. The sacks are made of jute, but fine mesh polyethylene is now available and this is very strong and durable. The sacks of concrete are usually laid to bond, similar to block walling.

10.2.4 Grout Injection of Pre-placed Aggregate

Graded aggregate, similar to that used for conventional concrete, is placed in forms or in a prepared excavation. Cement grout is injected at the bottom of the aggregate and rises upward, displacing the water. Proprietary equipment and techniques are required for this method, but when it is carried out properly, concrete of excellent quality results. The method can be used with advantage in situations where concrete placing would be difficult, for example in underpinning. But it has been used successfully where placing by tremie and bottom-opening skips was also suitable.

It is normal practice for the specialist contractors who carry out this work to use admixtures in the cement grout, such as pulverised fuel ash, in order to improve the flow characteristics of the grout. The mix porportions of the grout are usually about 1 part ordinary Portland cement to 1·5 to 2·0 parts of clean concreting sand, grade C or M (BS 882). The w/c ratio should not exceed 0·5. Quality control of the strength of the resulting concrete is difficult and special methods have to be used to obtain realistic test cubes.

Further details of this and other methods of underwater concreting are given in the Report of the Concrete Society previously referred to.

10.2.5 General Principles of Repair Below Low Water Level

As previously stated, the majority of repairs to marine structures involve patching or encasing reinforced concrete members such as piles, piers, etc. To do this below water level using cementitious materials (concrete or mortar) is very difficult. Each job presents special problems and the most satisfactory technique has to be worked out for each project. Cracks which have not adversely affected the strength of the member, and other minor damage, can be successfully repaired with epoxide resins and resin mortar. These resins are now formulated so that they can be applied under water. Figure 10.1 shows special equipment for spraying epoxide resin under water.

However, serious structural damage requires repair with concrete or gunite. The first step is to ascertain the extent of the damage. When this is below water level, closed circuit television is very useful in providing information for an initial appraisal of the problem. However, careful examination by divers is usually essential for an accurate assessment of the damage so that a realistic method of repair can be worked out.

If the structure has been in existence for any length of time, the concrete will be covered with marine growths (seaweed, barnacles, etc.). It will then be necessary to remove these growths completely before the full extent of the damage can be appreciated. Marine growth of all kinds can be quickly and completely removed by high velocity water jets. High velocity water, using nozzle pressures of about 400 atm, has been used for many years for cleaning ships' hulls and its use has now spread to the construction industry for cutting and cleaning concrete.

A typical method of repair to badly damaged piles below water level is to provide a reinforced gunite sleeve. The sleeve is formed in sections and lowered progressively until it reaches its final position; the annular space between the sleeve and the pile is then grouted in.

The author is indebted to the Cement Gun Company for information on an actual job carried out which is summarised below. The damaged members were precast prestressed tubular piles which supported a reinforced concrete platform some 28 m above the sea bed. After a careful investigation (which comprised both closed circuit television and examination by divers) of the damaged piles, it was decided that these would

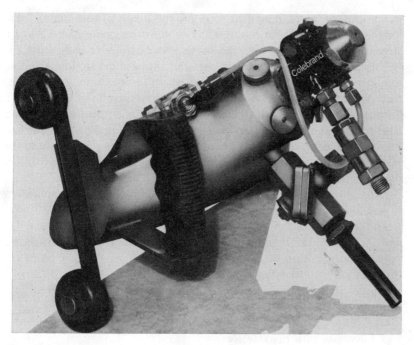

Fig. 10.1. Equipment for underwater application of resin coatings on concrete (courtesy: Colebrand Ltd).

be sleeved from above low water level down to about 1·5 m into the sea bed, which would have to be jetted out for the purpose. Above low water level the piles would be encased direct in reinforced gunite.

A loose sleeve of sheet steel was prepared and fixed in position around the pile above water level, the sleeve being separated from the pile itself by spacers. The shutter (sleeve) was coated with a release agent and wrapped in hessian to ensure a quick and clean release from the gunite shell. A reinforcing cage was then placed in position around the hessian wrapped shutter and high quality gunite applied to a total thickness of about 75 mm. This allowed about 20 mm of cover to the reinforcement on the inside and 35 mm on the outside. Experience has shown that with properly placed high quality gunite, cover of this thickness is adequate. The gunite was dense and impermeable and could be relied upon to have a compressive strength of not less than 45 N/mm². HAC was used for all the guniting work on this job. This enabled a complete unit to be gunned every 24 h.

The sleeves were gunited in sections, the longest being about 9·0 m long and weighing about 6 tonnes. As soon as the gunite had reached a strength of about 30 N/mm² the sleeve was carefully released from the shutter and lowered below water level, but with a short section including reinforcement projecting to connect with the next section, which was then constructed in the same way. In this way monolithic sleeves were formed of sufficient length to enclose the whole pile from below sea-bed level to just above low water level. As soon as a full length sleeve had been completed in this way, the annular space between the sleeve and the pile was grouted in with a specially formulated grout which had to displace the sea water and set under water.

Above low-water level the piles were repaired by the direct application of gunite after a reinforcing cage had been fixed around the pile.

The execution of the work was complicated by the tide range of 6·0 m and the strong, rather unpredictable tidal currents. The platform had to be kept in operation throughout the repair as it was connected to the main jetty by a 150 m walkway.

Figure 10.2 shows a reinforced gunite sleeve being lowered into position to below low water level.

10.2.6 Repairs with Epoxide Resins

Cases occur where some minor damage is caused to piles or other parts of a structure below low water level. In its initial state, the damage may not impair the strength or stability of the structure, but unless it is repaired, deterioration will be progressive and eventually a major and very costly repair, or replacement of the defective member, will be necessary.

The damage is usually in the form of cracks (sometimes of unknown origin), spalled arrises and small damaged areas of concrete. All these greatly increase the chance of sea water penetrating to and corroding the reinforcement, which is likely to result in large scale disintegration of the concrete member. In no circumstances is the old saying 'a stitch in time saves nine' more appropriate than the immediate repair of minor damage to a marine structure.

Regardless of the length of time a member has been immersed in the sea, general cleaning and preparation of the surface of the concrete is required. The amount of preparatory work will depend on site conditions and degree of damage. In warm, tropical waters marine growths appear and flourish much more quickly than in the cooler seas.

The type of repair dealt with in this section can now be carried out by

FIG. 10.2. Reinforced gunite sleeve for repair of jetty pile below low water level
(courtesy: Cement Gun Co. Ltd).

specialist firms using modified epoxide resins. These resins can be
formulated so as to have many of the most desirable characteristics to
suit site conditions. The resins will set and cure underwater and can be
applied by airless spray, brush or roller.

For the preparation of the concrete prior to the application of the
resin, high velocity water jets are likely to give the best results. If site
conditions and/or the size of the repair do not justify the use of the
necessary equipment, then general cleaning with wire brushes has to
suffice. As with all concrete repairs, the importance of careful and
thorough preparation of the damaged surface is essential. It is important
that the application of the resin should follow the cleaning operation as
rapidly as possible.

In some cases, such as estuaries, docks and harbours, the water may be so turbid that visibility is very poor indeed. This naturally makes repair very difficult, particularly the preparatory work. The resin, in the form of a putty, can be applied by hand ('gloved on') and repairs have been carried out successfully in this way.

Epoxide mortars, composed of the resin and clean, dry, carefully graded silica sand, can be used for repair of damaged arrisses and other defective areas. Generally the mortar should be applied in thin layers, each not exceeding about 4 mm thick. However, if thixotropic fillers are used in conjunction with the fine aggregate, this thickness can be appreciably increased; the actual thickness of each layer should be determined by trial application. In this way the mortar repair can be built up to almost any required thickness. Each layer has to 'set' before the next one is applied. The term 'set' means that the first layer is not disturbed by the application of the next layer. When the work is carried out below water level, this delay can be time consuming and expensive.

Where thicker layers are required, the gunite sleeve method can be used where this is suitable (e.g. for such members as piles). For more general application, a specially designed clamp-on shutter is appropriate.

The concrete mix has to be designed with a high slump so as to be virtually self-compacting, and for durability such a concrete will need to contain not less than 450 kg of cement/m^3; the w/c ratio should not exceed 0·40. Plasticisers can be used, but an excessive dosage must be avoided, as this can result in permanent loss of strength. Superplasticisers can be very useful for this type of work as they allow the use of a concrete with a low w/c ratio. In this way, repairs to concrete, with concrete, can be carried out below water level. This has many obvious advantages for work below low tide level and between the tides. Some information on superplasticised concrete is given in Chapter 1, and in Section 10.3 below.

10.3 REPAIRS BETWEEN THE TIDES AND ABOVE HIGH WATER LEVEL

The section between the average levels of low tide and high tide is in reality the lower part of the 'splash zone'. This zone is generally considered to be the one which is subjected to the most severe conditions of exposure. It is in this zone that the greatest fluctuations in moisture content and temperature of the concrete occur. The period during which the concrete is submerged varies. The tide cycle around the coast of the UK is about 12 h from high tide to high tide. Clearly, concrete just below high tide level is totally submerged for the shortest period while that just above low tide is submerged for the longest period. When the concrete is above water level it may be subjected to wind driven spray as well as sand and shingle thrown against it by the waves with very considerable force. The problem of physical abrasion is discussed in the next section.

In regions such as the Persian Gulf and the Red Sea, the sun temperature can reach 60–65°C and when concrete at or near this temperature is covered with highly saline spray, the spray is likely to be readily absorbed into the surface layers. At night the temperature drops rapidly, and so these changes in temperature plus absorption of salt create conditions likely to result in disintegration of the surface layers of concrete. A similar wetting and drying, heating and cooling cycle occurs in the temperate zones, but within a much smaller range.

At the other extreme of climate, in the far North and South, the sea freezes, also the spray, together with the moisture trapped in the surface layers of the concrete. The expansion of the absorbed moisture can create stresses sufficient to crack and spall the concrete. To mitigate the effect of freezing temperatures, air-entrained concrete is often used. The entrainment of about $4\frac{1}{2}\%$ of air alters the pore structure of the concrete and this is very effective in preventing scaling and spalling under freezing conditions.

The author has not seen any recorded use of air-entrained concrete for marine work other than in very cold climates, but there would appear to be some advantage in its use, particularly where the concrete may be vulnerable to disintegration by the build-up of salt in the surface layers. However, the basic requirements for high quality concrete which have been set out earlier in this chapter should be followed. It can be difficult to effectively air-entrain cement-rich mixes.

Repairs to concrete marine structures in the tidal zone are always a race against time. In no other section of the work is careful and detailed preparation and organisation more necessary. Everything required for the work, materials, equipment and labour, must be so prepared in advance that as soon as the tide turns, the next section of the work can be put in hand without delay. The critical section is near low water where the turn-round time is shortest. The usual tide period in the UK is six hours from high to low and six hours low to high. This means that the concrete must be protected to prevent damage to the surface by the incoming tide, which always involves a certain amount of wave action and often fairly strong currents as well. A storm can be disastrous.

For the protection against tidal action (but not storms) of horizontal and sloping surfaces, the use of jute sacks stiffened by an application of cement grout made with about 50% OPC and 50% HAC, can be very effective. This mixture generally produces a flash set. The exact proportions must be found by trial and error as the speed of set will depend on the chemical composition of the two cements actually used. The sacks must be weighted down with large stones or similar to prevent them being moved by the incoming tide. An alternative to the use of jute sacks is to spray the cement mixture direct onto the plastic concrete, followed immediately with a light water spray. This should set almost instantaneously, forming a crust. Further protection in the form of jute sacks, fine mesh polyethylene or boarding, can then be laid on this protective 'crust'.

Apart from the requirement to provide this initial protection the repair work proceeds in the usual way, but with the utmost dispatch, using the basic principles of mix proportions, cover to reinforcement, and compaction, all as discussed earlier in this chapter.

Where formwork has to be used, as for columns, beams and suspended slabs, then this in itself provides protection until the concrete is mature enough to resist physical damage. The period during which the formwork should remain in position may therefore by appreciably longer than would be required for a land-based structure. It should be remembered that a storm can blow up very quickly, and concrete which has not reached an adequate strength can be badly damaged by the pounding of waves, sand and shingle. In these circumstances it is not practical nor desirable to lay down in advance how long the formwork should be kept in position. The specification should contain a flexible clause clearly allowing the Resident Engineer to decide when the formwork can be struck. To be fair to the contractor and thus allow him to price the job realistically, the Bill should contain a provisional item for keeping the formwork in position beyond a certain time.

There is a great difference in the rate of hardening of Portland cements and HAC. If HAC is used then it is unlikely that the formwork need be kept in position longer than 24 h after completion of casting. With ordinary or rapid hardening Portland cement it may not be wise to remove the formwork for at least 7 days or even longer. However, the cost of HAC is about four times that of ordinary Portland cement.

Mention has been made previously that normal Atlantic water will not be aggressive chemically to good quality concrete made with Portland cement, and the same applies to HAC.

In estuaries, harbours, and other sheltered positions, the chemical characteristics of the sea water may be significantly different to those in the open sea. It is therefore advisable, when considering repairs to marine structures in such areas, to check whether or not the deterioration of the existing concrete is due, or partly due, to chemical attack caused by unsuspected concentrations of aggressive chemicals in the water or in the silt and mud through which the piles, etc., pass. The lower reaches of many rivers are grossly contaminated with industrial waste. Samples of the water and of the mud, etc., which will be in contact with the concrete should be taken at intervals over as long a period as possible so as to provide maximum information on the chemical composition and its probable variation.

In some cases the damage is so extensive that the repair necessitates the placing of an appreciable thickness of concrete to the sides and soffits of beams and slabs, and to columns. Crack injection is often required in order to reduce ingress of sea water in members which are otherwise undamaged.

When placing concrete, site conditions may be such that ultra-high workability is required. This is known as 'flowing' concrete and is obtained by the addition of a superplasticiser, but with the water/cement ratio maintained at a maximum of 0·40. It is likely that a 'flow' of 550–650 mm will be needed and the method of test for this degree of workability is laid down in BS 1881—Part 105, Testing Concrete: Methods of Determination of Flow.

It is essential that the concrete should not segregate at any stage from batching to placing and compacting; also, the concrete must maintain its maximum workability during placing and compacting. Trial mixes are essential, and must be carried out using the cement and fine and coarse aggregates which will be used on site.

The two types of superplasticiser which are suitable for marine work are:

(a) Sulphonated melamine formaldehyde.
(b) Sulphonated naphthalene formaldehyde.

Regarding curing, it may be thought that the high humidity associated with a marine environment would render curing less important. This is definitely not so, as wind can cause severe plastic cracking even when the humidity is high. Therefore the author recommends the following procedure:

> For Concrete: Immediately formwork is removed a curing membrane should be applied by airless spray so as to cover the whole of the exposed concrete within three hours of formwork stripping.
>
> The top exposed surface of concrete cast within formwork and the surface of any horizontally cast concrete (slabs, etc.) must be completely covered with the curing membrane within 30 min of completion of placing and compaction.
>
> For Gunite and Mortar: All gunite and mortar should be completely covered with the curing membrane within 30 min of completion of placing and finishing.

There are many advantages in selecting a curing membrane which will also serve as a coating to seal the surface of both new and existing work in the long term. This type of membrane is usually based on epoxide resin and is applied as a primer and finishing coat, both by airless spray. It follows that the primer must be suitable for application to 'green' concrete and mortar. The minimum thickness for primer and finish should be 0·5 mm.

Essential characteristics of the membrane are excellent bond to the substrate and durability in a marine environment. The test procedures set out in BS 3900—Tests for Paints, Part E6: Cross-cut Test, and Part E10: Test for Adhesion, can be usefully applied to the membrane, but only after it has fully cured, which usually requires at least 7 days. Trial areas should be carried out and the results, assuming these are satisfactory, can be used as a basis for comparison with the actual coating work on the structure. The aim should be to reach a classification not lower than 1 for the cross-cut test (Table 1, Part E6). There must be a reasonable tolerance, and the author suggests that not more than one test out of each six should fall below classification 2.

For the adhesion test (Part E10), the coating should not fail in adhesion at the interface between the primer and first coat; if failure of adhesion does occur it should be within the surface layers of the concrete. The author considers that the adhesion test is mainly suitable for the pre-work test specimens, but the cross-cut test can be applied to the completed work as well.

Prior to the application of the membrane to existing concrete, the surface should be thoroughly water jetted to remove all salt and other contamination.

Crack injection. It is usual to find that concrete marine structures which have been physically damaged by the sea or impact from vessels, also contain numerous cracks. In addition, fine cracks often occur in new concrete placed as part of the repair, particularly when this is in appreciable areas. A reasonable criterion is that all cracks which are 0·20 mm or wider should be injected with a polymer resin. This 0·2 mm would be the width on the surface, and the resin selected should be capable of penetrating cracks down a width of 0·05 mm, as cracks usually decrease in width with increase in depth.

The object of the injection is to fill the cracks as completely as possible with resin which must be durable in a marine environment, particularly in the splash zone. The contractor should be required to demonstrate to the satisfaction of the Engineer that:

(a) He has had previous experience in the injection of cracks in the splash zone of marine structures, and
(b) he has successfully injected cracks down to a width of 0·05 mm.

The contractor should also be required to carry out on the structure trial crack injection, which should be followed by taking 50 mm cores through the injected concrete to show the resin penetration actually achieved.

It is advisable for the contractor to be required to provide information on the type of resin he proposes to use, and its viscosity and percentage of filler. The Engineer should check these from time to time during the injection process.

10.4 REPAIRS TO DAMAGE CAUSED BY ABRASION BY SAND AND SHINGLE

This type of damage is mainly confined to sea walls, groynes, promenades and piers. Figure 10.3 shows severe abrasion to the base of a pier.

Since about 1967, special attention has been paid to this problem by the Sea Action Committee of the Institution of Civil Engineers.

In considering abrasion of marine structures, the author feels that reference can usefully be made to the information available on the abrasion of concrete floors of industrial buildings. It is admitted that environmental conditions are quite different, but a finish which will stand up to steel wheeled trolleys may give good results for wave-driven sand and shingle.

It is known from experience with industrial floors that concrete with a

Fig. 10.3. Severe abrasion by sand and shingle to base of concrete pier (courtesy: Cement & Concrete Association).

high cement content and low w/c ratio has considerably higher abrasion resistance than a leaner concrete using the same aggregate. It has been found cheaper to increase the cement content, lower the w/c ratio and use a plasticiser, than to import a harder aggregate from a long distance. The effect of the hardness of the aggregate on the abrasion resistance of the concrete floor is less than that of the overall quality of the concrete.

An interesting example of the abrasive resistance of high quality cement-rich concrete is the apron to the promenade at Littlestone, which was reconstructed by the Kent River Authority between 1960 and 1966 using ragstone slabs set in fine concrete. The concrete had mix proportions of 1 part sulphate-resisting Portland cement to $2\frac{1}{2}$ parts of flint gravel including sand, with 4% air entrainment and 2% calcium chloride added to speed-up the setting and hardening. The w/c is not known, but the cement content of the mix was about $620 \, kg/m^3$. The concrete has stood up very well to abrasion, in fact better than the ragstone slabs. There was no reinforcement in the concrete.

At the present time it is not possible to make specific recommendations based on research or controlled experiments. Therefore when dealing with repair work it is necessary to base the proposals on experience gained in similar work and to adapt techniques which have been found satisfactory in other types of work, as far as this is practical.

The author therefore suggests the following:

(a) The concrete mix to contain a minimum of $400 \, kg$ cement/m^3 concrete.

(b) For Portland and HAC cements the w/c ratio should not exceed 0·4.

(c) The aggregate should be as abrasion resistant as possible; flint gravel and crushed granite are generally more resistant to abrasion than limestones and sandstones. If relatively small quantities of concrete are required, then it may be worthwhile considering the use of HAC with a special aggregate such as Alag, which is crushed and graded HAC clinker. Alag should only be used with HAC.

(d) Special precautions should be taken to protect the concrete (or gunite if this is used for the repair) from being damaged.

10.5 MARINE GROWTHS ON CONCRETE

Even in cold and temperate seas, marine growths appear very rapidly on concrete structures. In warm tropical waters, this growth is extremely rapid. In both cases, it is very difficult to remove the seaweed, barnacles, etc. Questions are sometimes asked as to the long term effect of these growths on the durability of concrete. The author has found no published information which suggests that the durability of good quality concrete is adversely affected by seaweed, barnacles and the many thousands of other marine growths. There is probably a small removal of lime from the surface layers, but that is all; with adequate cover to the reinforcement of dense impermeable concrete, the effect is insignificant.

It is often necessary to remove these growths from slipways and steps, promenades and aprons to sea walls, because they make the surface of the concrete very slippery. In such cases, high velocity water jets are likely to give the best results.

The use of poisons mixed into the concrete or applied to the surface is dangerous in locations where people will walk or otherwise come into contact with the concrete. These antifouling preparations only inhibit the growth of marine organisms for a short period and therefore require regular renewal.

However, marine growths can seriously interfere with the carrying capacity of sea water intakes to power stations and research has been carried out to find some permanent technique for preventing these growths. A detailed report on this work has been published by the Civil Engineering Laboratory at the Naval Construction Battalion Centre in California; details of this report are given in the Bibliography at the end of this chapter. It was found that a durable antifouling concrete could be made with a porous expanded shale aggregate which had been impregnated with creosote and certain other toxic chemicals. The resulting concrete remained essentially free of marine growths for a period of four years. This special concrete was of medium strength, about 25 N/mm^2. The samples were exposed to marine conditions near the surface of the ocean and at a depth of about 35 m, off the coast of Cuba and California.

10.6 REPAIRS TO FIRE DAMAGED MARINE STRUCTURES

It is seldom that marine structures are damaged by fire, but when this does occur it is usually a hydrocarbon fire such as that caused by an oil tanker catching fire when alongside a terminal structure. The effect of such a fire can be very severe indeed. Figure 10.4 shows one small section of what was overall very extensive damage. Spalling of concrete extended more than 300 mm behind the 32 mm diameter reinforcing bars.

The cost of remedial work can be extremely high, especially if the terminal is not connected to the land. High quality, superplasticised concrete and high quality gunite should be used, as appropriate, for the repair of the various members.

If the repair is delayed for any reason, there is likely to be rapid corrosion of exposed reinforcement in the splash zone thus increasing the amount of steel which has to be replaced.

Special precautions must be taken in curing newly placed concrete and gunite, as sea 'breezes' can cause severe plastic cracking. The use of a curing membrane which can act as a primer or base coat for a subsequent protective/sealing coat to seal in the entire exposed surfaces of all repaired members, is strongly recommended.

FIG. 10.4. Severe damage caused by hydrocarbon fire to marine terminal structure (courtesy: Burks, Green & Partners, Consulting Engineers).

BIBLIOGRAPHY

AMERICAN CONCRETE INSTITUTE, *Guide to durable concrete* (reaffirmed 1982), ACI Committee 201; ref. 201.2R–77, 1977, p. 37.

AMERICAN CONCRETE INSTITUTE, *Erosion of concrete in hydraulic structures*, ACI Committee 210; ref. 210R–55 (reaffirmed 1979), p. 10.

AMERICAN CONCRETE INSTITUTE, *Performance of concrete in marine environment*, ref. SP–65, 1980, p. 640.

BRITISH STANDARDS INSTITUTION, BS 6349, *Code of Practice for maritime structures*, Part 1, 1984, *General Criteria*.

BRITISH STANDARDS INSTITUTION, BS 8110, *The structural use of concrete*, Parts 1 and 2: 1985 (replaces CP 110).

BUILDING RESEARCH ESTABLISHMENT, *The durability of reinforced concrete in sea water*, 20th Report of the Sea Action Committee of the Institution of Civil Engineers; National Building Studies, Research Paper No. 30, 1960, HMSO, London, p. 42.

CONCRETE SOCIETY, *Underwater concreting*, Tech. Report 52.018, Cement & Concrete Association, London, 1975.

DEPARTMENT OF INDUSTRY, *Concrete in the Oceans*, Reports published by Cement & Concrete Association, London.

GJORV, O. E., Long-term durability of concrete in sea-water. *J. Am. Conc. Inst.*, Jan. 1971, 60–77.

Lea, F. M., *The Chemistry of Cement and Concrete*, 3rd ed., Edward Arnold, London, 1970, p. 727.

Muraoka, J. S. and Vind, H. P., *Anti-fouling marine concrete*, Civil Engineering Lab. Naval Construction Battalion Centre, Calif. USA, May 1975, p. 22.

Appendix

Testing Concrete and Mortar in Existing Structures

1. INTRODUCTION

There are, essentially, two types of testing:

(a) Testing involving the taking of samples from the concrete and/or mortar.
(b) What is known as 'non-destructive' testing in which samples are not taken.

When samples are taken, the concrete/mortar has to be made good and the samples are normally destroyed by the testing process. The samples are taken to establish certain characteristics, e.g. compressive strength, grading of aggregates, type of cement, mix proportions, percentage of chlorides, sulphates, etc.

The range of 'non-destructive' tests has increased considerably in recent years, and now includes covermeter surveys, ultrasonic pulse velocity tests, Schmidt hammer tests, loading tests, gamma radiography, half-cell surveys.

A great deal has been written and published on all types and methods of test; a small selection of the available literature is given in the Bibliography at the end of this Appendix.

It is important to remember that all types of testing require experienced and intelligent interpretation.

2. TESTING INVOLVING TAKING SAMPLES

The main tests are to determine strength (usually compressive strength by means of cores), and chemical analysis, microscopic and petrographic

283

examinations, and visual inspection of the samples, and the concrete/ mortar surrounding the extraction of the core or sample.

The field for argument and disagreement usually revolves around the assessment of compressive strength and the amount of sampling required to establish the overall condition of the concrete/mortar (strength and mix proportions). At the time of writing this book there were no published recommendations for the number of cores required to enable a reasonable assessment to be made of the insitu strength of the concrete, nor on the amount of hardened concrete (number of samples) which are needed to determine the 'average' cement content (mix proportions).

Sampling of mortar screeds is covered by BS 4551. The sampling rate at first sight appears large, one subsample of 100 g per 10 m², and one main sample to consist of ten sub-samples. However, for a screed or topping 40 mm thick, the one 100 g sample per 10 m² represents 100 g in about 700 kg of mortar, i.e. one part in 7000.

For concrete, the author feels that the recommendations for rates of sampling given in Table 7 of BS 5328:1981 could be followed, provided that each 'sample' consists of two cores. So that for a floor slab containing say 500 m³, there would be 20 cores $(500 \div 50) \times 2$. For determination of mix proportions, 50 mm diameter cores are adequate.

For determination of chloride content in hardened concrete or mortar, the recommendations in Building Research Establishment Information Sheet IS.13/77 should be followed. Sulphate content can be determined from the remains of crushed cores, or other samples.

For chemical analysis to determine cement content, it is relevant to note that the actual amount of concrete/mortar which is analysed is very small indeed, about 10 g. In assessing the analysis results, allowance must be made for unavoidable errors and BS 5328 suggests an allowance of plus or minus 10%, but some engineers work to ± 30 kg/m³.

Petrographical tests are carried out much less frequently and when they are, it is usually in connection with suspected alkali–aggregate reaction in the concrete.

Microscopic examination is used to determine, among other things, whether sulphate-resisting Portland cement has been used, and this test requires the careful preparation of thin sections.

The test for depth of carbonation is carried out on cores and/or on the concrete exposed by the taking of samples. It is a very useful and easily applied test as it indicates by a distinctive colour change the advance towards the reinforcement of the carbonation front.

3. NON-DESTRUCTIVE TESTING

It can be reasonably claimed that all forms of non-destructive testing are indicative only and should not be considered as giving absolute information.

Electromagnetic covermeter surveys can under favourable conditions be considered as accurate on an average site, within about ±5 mm.

The Schmidt Rebound Hammer measures the surface hardness of the concrete, but the readings can be used to indicate the likely strength of the concrete; the results should be used in conjunction with other relevant information.

Ultrasonic pulse velocity tests are also very useful in assessing concrete strength and investigating the likely presence of cracks and voids below the surface.

Neither the Schmidt Hammer tests nor UPV tests should be used on their own to condemn concrete.

There are other tests for assessing the compressive strength of insitu hardened concrete in structures; these include an internal fracture test developed by the Building Research Establishment, and the LOK and CAPO tests developed in Scandanavia. Papers and reports describing these tests are included in the Bibliography at the end of this Appendix.

Reference was made in Chapter 3 to the use of the copper–copper sulphate half-cell for detecting corrosion of reinforcement embedded in concrete. This is a very useful technique, developed in the US and for which there is an ASTM Standard (No. C 876–80).

Gamma radiography is used to determine the location of reinforcement and its size. It is expensive, and strict health and safety precautions are needed in its use.

It has been pointed out in this book that the main cause of deterioration of reinforced concrete is the corrosion of reinforcement, and this is directly influenced by the quality and thickness of the concrete cover. In this case, 'quality' really means 'low permeability'. The author therefore feels it is unfortunate that there is still no practical and satisfactory site test for permeability. The ISAT test (initial surface absorption test) was originally used for precast concrete products and its use on site presents many practical difficulties. The Building Research Establishment developed a test using a hole drilled into the concrete, but there is little information available on its use in the field.

There are two types of load test:

1. Service load test, to check compliance with the design and specification.
2. Test to assess safety of the structure.

It is usually the second type of test which is used in investigations involving repair and strengthening of concrete structures. The object is to ascertain if the structure or a specific part of the structure, can carry the required service load with an adequate margin of safety.

It is important to observe and record the behaviour of the structure during the test, i.e. deflexions, recovery on removal of load, and the development of cracks (their location, direction and width).

Load tests must be carried out with great care and by an experienced firm, with a qualified and experienced engineer continuously on site during the test. Provision must be made to contain any unexpected collapse.

Load tests are very expensive in time and money; the client often views it like an insurance policy—if it gives satisfactory results and shows the structure is adequate he feels he has wasted his money!

The author again emphasises that non-destructive testing is extremely useful, but the results need careful and experienced interpretation in conjunction with other relevant information.

BIBLIOGRAPHY

AMERICAN CONCRETE INSTITUTE, *Concrete core tests* (Bibliography No. 13); ACI Committee 214; ref. B–13, 1979, p. 30.
AMERICAN CONCRETE INSTITUTE, *Strength evaluation of existing concrete buildings* (revised 1982); ACI Committee 437; ref. 437. R–67, 1982, p. 6.
AMERICAN SOCIETY FOR TESTING AND MATERIALS, C 900–82—*Standard test method for pull-out strength of hardened concrete.*
AMERICAN SOCIETY FOR TESTING AND MATERIALS, C 876–80—*Standard test method for half cell potentials of reinforcing steel in concrete.*
BRITISH STANDARDS INSTITUTION, BS 1881—*Testing concrete.*
BRITISH STANDARDS INSTITUTION, BS 4551—*Methods of testing mortars, screeds and plasters.*
BRITISH STANDARDS INSTITUTION, BS 6089—*Assessment of concrete strength in existing structures.*
BRITISH STANDARDS INSTITUTION, BS 4408—*Recommendations for non-destructive methods of test for concrete* (under revision).
BRITISH STANDARDS INSTITUTION, BS 5328—*Methods of specifying concrete including ready-mixed concrete* (under revision).
BRITISH STANDARDS INSTITUTION, BS 8110—*Structural use of concrete*, Parts 1 and 2 (this replaces CP 110).

BRITISH STANDARDS INSTITUTION, BS 3683—*Glossary of terms used in non-destructive testing.*

BUCKLEY, J. A., The variability of pull-out tests and in-place concrete strength. *Concrete International,* April 1982, p. 8.

CHABOWSKI, A. J. AND BRYDEN SMITH, D. W., *Internal fracture testing of insitu concrete—a method of assessing compressive strength.* Building Research Establishment Information Paper IP.22/80, October 1980, p. 4.

INSTITUTION OF STRUCTURAL ENGINEERS, *Appraisal of existing structures,* Report, July 1980, p. 60.

JONES, D. S. AND OLIVER, C. W., The practical aspects of load testing, *Structural Engineer,* Dec. 1978, **56A** (12), 353–6.

KRENCHAL, H. AND PETERSEN, C. G., *Insitu testing with LOK-test; ten years experience.* Paper at Research Session at International Conference on Insitu Non-destructive Testing of Concrete, Ottawa, October 1984, p. 24.

MALHOTRA, V. M., *Testing hardened concrete: non-destructive methods* (Monograph No. 9), ACI: ref. M–9, 1976, p. 204.

MENZIES, J. B., Load testing of concrete building structures. *Structural Engineer,* Dec. 1978, **56A** (12), 347–53.

PETERSEN, C. G., LOK-test and CAPO-test development and their applications *Proc. ICE,* Part 1, 1984, 76, May, 539–49.

STRATFULL, R. F., *Half-cell potentials and the corrosion of steel in concrete,* US Highway Research Record 1973, 12–21.

TOMSETT, H. N., The practical use of upv measurements in the assessment of concrete quality, *Mag. Conc. Res.,* **132** (110), March 1980, 7–16.

VASSIE, P. R. W., *Evaluation of techniques for investigating the corrosion of steel in concrete;* Transport and Road Research Lab., Crowthorne, p. 22.

Index

Abrasion, effect on concrete, 49
Abrasion resistance, testing of, 157
Abrasion-resistant finishes
  floors, 158
  marine structures, 277–9
  pipelines, 256, 257–8
Absorption properties (of concrete),
  36, 87–8
Accelerators, 14
Access equipment (for investigations),
  65, 68, 189, 190
Acids, effects on concrete, 38–40, 43,
  233
Acrylic emulsions, as bonding aid, 20,
  100, 158, 208, 236
Acrylic resins, 26, 29
Actual strength (of concrete), 86
Additives, 13
Admixtures, 13–20
  accelerators, 14
  air-entraining agents, 15
  condensed silica fume, 19–20, 167,
    169
  definition of, 13
  marine repair work, for, 265
  plasticisers, 16–18
  reason for use, 14
  retarders, 15
Aggregates, 11–13
  abrasion resistance (of concrete)
    affected by, 49, 279
  characteristics listed, 11–12
  marine repair work, 264
  slip resistance of surfaces, and, 49,
    173

Aggregates—*contd.*
  temperature effects on, 138–9
Agrément Certificates, bridge deck
  membranes, 181, 207
Air-entrained concrete, 49, 115, 265,
  273
Air-entraining agents, 15, 115
Airfield runways, de-icing of, 45, 204
Alag aggregate, 279
Alkali–aggregate reaction (AAR), 11,
  45–8
  testing for, 284
Alkali–silica reaction (ASR), 46–8,
  210–11
  bridges, in, xiii, 202, 210–11
  cracking caused by, 47–8, 76–9, 210
  methods of dealing with, 211
  visible symptoms of, 47, 210
Alkalinity (of concrete), 38, 46, 54,
  208, 246, 247
Aluminium
  chemical attack on, 10
  protection of, 56
  temperature effects on, 137
American Concrete Institute (ACI)
  chemical attack data, 40, 41–2
  shotcreting code of practice, 131
  silo design recommendations, 133
American Society for Testing and
  Materials (ASTM) standards,
  7, 9, 20
Ammonium-based fertilisers, 20, 40,
  41, 169
Ammonium compounds, effects on
  concrete, 40–1